Advances in Horticulture Science Research

Chandralekha Arora

Oxford Book Company
Jaipur ▪ New Delhi

Published By

Oxford Book Company

Regd. Off. "Bony Residency", Gate No. 2
Opp. Tilak Public School, Vishweswariya Nagar,
Gopalpura Road, Jaipur - 302018, India
Ph: 0141 - 2761280, Telefax: 2761381
email - oxfordbookjpr@gmail.com
website - oxfordbookcompany.com

Delhi Office: 102, 1st Floor, Satyam House,
4327/3, Ansari Road, Darya Ganj,
New Delhi - 110002, India
Ph: 011-45652440

Distributed By
KEYA INTERNATIONAL PRESS LLP
E-mail: keyainternationalpressllp@gmail.com

Published in India, 2024

ISBN : 9788189473808

Price: ₹ 1925.00

Typeset by Shivangi Computers, Jaipur

Printed and bound in India by Thomson Press Ltd.,
New Delhi

Preface

Horticulture is the area of agriculture that deals with garden cultivation: the science and art of growing fruits, vegetables, ornamental trees bushes, and flowers. Along with forestry, horticulture and agronomy constitute the three broad branches of agricultural science and practice. Today as the demand for exotic fruits and flowers increases, and new markets open up for horticultural products, research in this science has also taken great strides. The primary aim of horticulture science is to develop plants of the highest quality that offer the promise of high yields. The advancement in technology thereby has enabled this, which augurs well for the future of horticulture.

The present text, designed as a manual for horticulturists, brings within purview the contemporary advances in horticulture research. The emphasis is on understanding how newer techniques and technology has enabled better propagation of crop plants, with improved quantity and quality, as well as resistance to pests and disease. The focus is on delineating the principles concepts and practices which give shape to horticulture in general, and how current trends and developments are bringing about a change for the better. In addition; it also provides insight into the future of horticulture, paying particular attention to the sciences of genetic engineering and biotechnology, which have revolutionised many other scientific branches.

Chandralekha Arora

Contents

1

Modern Trends in Horticulture Research

Horticulture playing an important role in the modern agricultural researches. It is indeed not easy to delineate horticulture, to hit on a definition that adequately expresses the modern concept of the word. One attempts to trace its boundaries only to discover overlappings in other fields, controversial spheres of influence. Yet there is a popular notion of horticulture which brings people to flower shows, to cherry blossom festivals, to the market place. That practical horticulture and research are inexorably bound together is a fact that cannot be ignored.

It must be obvious to anyone that horticulture has to do with certain kinds of plants or crops which have through long custom been associated with it. Furthermore horticulture represents a special type of viewpoint. This latter is rather intangible but none the less true. In the beginning horticulture referred to a garden, that is, to an intensive type of plant growing in contrast to the culture of field crops. This meant fruits, flowers, vegetables, and ornamental plants. It represented a certain refinement of agriculture which came with leisure or was associated with the home itself.

As plantings were expanded for commercial purposes the old idea of interest in the individual plant was retained. Whether apple, potato, or rose farming is still horticulture might be debated by some. But certainly horticulture has emerged from a garden into a vast enterprise and includes such features as nursery practice, the growing of vegetable and flower seeds by the thousands of acres, the processing of canned and frozen foods, the manufacture of spray materials and fertilisers, as well as the growing of fruits, flowers, vegetables, and ornamental plants.

Finally, it must be pointed out that horticulture is both an art and a science. It was an art long before there was any science or any agricultural colleges, and the art will continue whether there is a science or not. It is practiced by the untaught and by those of the greatest learning and culture. That is, horticulture, the art, does not belong to horticultural departments, neither is it destroyed no matter into whose hands it may find itself for a time. It retains its distinctive features, no matter into what university melting pot it may be thrown.

PLANT'S ENVIRONMENT

If we consider ecology as that branch of biology which deals with the mutual relations between organisms and their environment, then horticulture makes the widest use of it. Ordinarily we do not call the studies by this name, and perhaps a new term will be introduced which will refer more particularly to applied or artificial relationships.

Some reference must, therefore, be made to the manipulation of environmental factors, particularly those that are within the hands of the operator. As Livingston says: In all our experiments the environment lurks behind the scenes, as it were, ever ready to baffle our efforts and always calling for very watchful alertness on our part.The grower of crops must deal with the environment below and above ground. As he changes the environment by cultural practices, whether it be tillage, irrigation, pruning, or spraying, he to some degree affects the plant behavior.

Temperature.—It is only in the greenhouse and in the refrigerated storage that much can be done about temperature. True, orchard heating is practiced in some sections, mulching is used in part to prevent great fluctuations in soil temperature, but in the main, the horticulturist must select varieties and cultural practices that give the least trouble. Temperature is, however, receiving considerable study and most marked effects are obtained. That temperature has an influence on growth, anatomy, and the metabolism of the plant is manifest.

An interesting and illuminating case of the effect of temperature upon the vegetative and reproductive phases of development is that reported for cabbage. Premature seeding or "bolting" of cabbage is of common occurrence. The cause was usually assigned to poor seed although it was known that early sowing of seed predisposed the crop to this trouble.

Work conducted under the direction of Thompsonat Cornell showed that "mature plants taken up in the fall before they were subjected to low temperature did not seed when grown subsequently in a warm greenhouse (60° to 70° F.), while those grown in a cool greenhouse (50° to 60° F.) produced flowers and seed. The plants grown continuously in the warm house produced three heads in succession during the first year. One of the plants was kept in the warm house for two years after it was brought in from the field in October, 1925, and this plant produced six heads. After the sixth head developed and the plant was about three years old it was moved to a cool house and within a few months it flowered. The plants used represented a pure-line strain so that all of them had the same genetic constitution."

The opposite of this situation obtains with lettuce, high temperatures favoring premature seeding. In none of these cases has length of day been a factor. With greenhouse plants grown for their flowers, equally interesting results have been secured. For instance, stocks, which also belong to the Criciferae, respond very similarly to cabbage, but since it is the flowers that are wanted the early "bolting" is desirable, and subjecting the seedling plants to low temperature may now be practiced.

Bulbous plants also are treated to force early flowering. Daffodils and bulbous iris subjected to 50° F. before forcing will bloom about four weeks early. If tulips are first preheated at a low temperature of 75° to 80° F. and then subjected to a low temperature they can be advanced considerably in time of bloom. This work has been done in Holland, England, and America. Darwin tulips can be secured in this manner in November instead of February as had been customary.

Lilies do not initiate flower primordia in the bulb stage as do other bulbs; hence it is possible to subject the young plants when a few inches high to a temperature of 50° for about two weeks. Such treatment will increase the number of buds per stock.

If gladiolus corms are placed in a temperature of 75° to 80° for a month before forcing, that is, preheated, they will flower about a month early. Equally striking anatomical differences may be noted at different temperatures.

Water—Water is such a commonplace that one would not expect to pay attention to it unless there was a great surplus or deficiency. Yet a successful greenhouseman will tell you that one of the most difficult operations to teach to a beginner is that of watering the plants. The desired amount varies with the kind of crop, the weather, the season of the year, the stage of development of the plants, and the nature of the soil. Many a crop is ruined by the misuse of water.

An interesting case of research in regulating water supply is provided in studies with tomatoes. Limited watering which allows the soil to be alternately wet and dry results in a large percentage of the crop (more than 75%) being affected with blossom-end rot, whereas similar plants which are, kept well supplied at all times show practically none of this trouble. In fact, this simple treatment has practically eliminated it from greenhouses.

Light—Following the classic work of Garner and Allard on the effect of length of day on plants, the horticulturist has made some practical use of this finding, notably with greenhouse crops. The chrysanthemum, being a short-day plant, was particularly amenable to treatment. Laurie and his associatesused black cloth to cover the beds in the greenhouse from 5:00 P. M. until 7:00 A. M. each day for a period of 30 to 35 days beginning July 15. This treatment has advanced the crop a month or more. On the other hand using electric lights to lengthen the day delayed flowering so that the mum season is now six instead of the usual three months.

Similarly, asters may be brought into bloom earlier by shortening the daylight period.

Just as shortening the day has induced early flowering of certain plants, so lengthening the day has had equally beneficial effects with certain other commercially grown flowering plants. The additional light should be applied from 5:00 to 10:00 P. M. or from 2:00 A. M. to 7:00 A. M. during fall and winter. Particularly responsive are Lilium longiflorum, annual chrysanthemum, cineraris, Gypsophila, and stocks.

Atmosphere—It is not easy to regulate the atmosphere about plants and perhaps as conspicuous an example of efforts in this direction as might be cited is the use of modified atmospheres for the storage of fruit, or gas storage, as it is called in England. Trials have been made for years on the effect of artificial atmospheres in the storage and transit of fruits. As early as 1821 Berard was presented with a medal by the French Academy of Science for his paper on the storage of fruit in atmospheres devoid of oxygen, but it is only within the past few years that interest in the subject has developed from a commercial standpoint.

Smock has described this type of storage and the results procured by it. He points out that respiration is diminished in intensity not only by lowering the temperature but also by reducing the amount of oxygen or allowing CO_2 to accumulate. Yellow Newtown apples held in CO_2 concentration of 10 per cent at a temperature of 45° F. were as green and firm as those held in air temperatures of 32° and 36°. Northwestern Greening apples after 30 weeks in 5 per cent CO_2 and 2.5 per cent oxygen (normal is 21 per cent O_2) at 40° F. were still as green as when first stored. Comparable fruit stored in normal air at 40° for the same period was completely broken down and worthless. Similar fruit stored at 36° was wholly unmarketable after 30 weeks because of softening, yellowing, and 100 per cent brown core. Excellent results were secured with McIntosh and other kinds of fruit.

The Soil—Probably the most conspicuous to those of us who have been associated with horticulture for a good many years is that of soilless culture or hydroponics, as it has come to be known in some quarters. The very thought of growing crop plants to maturity without soil, without organic matter, is rank heresy. How extensive it will become cannot be said, but the fact is that better quality flowers are being grown in a medium of gravel, cinders, or a similar inert material than in soil. In some cases the yields are also higher. For instance, 35 rose flowers are being produced per plant in gravel culture and 20 is the usual number in soil. Sweet peas average about 50 per cent more in gravel. Carnations produce no more cut flowers in gravel, but the blooms bring a premium on the market because of their quality. The same may be said of snapdragons, calendulas, stocks, lilies, daffodils, iris, and many annuals.

Among the vegetables, tomatoes are having a wider trial than others and there is some reason to suppose that they may be grown profitably without soil. Among the advantages of this system of culture are better aeration, uniform moisture conditions, better control of soil-borne diseases, a better balanced condition of the nutrients, and control of the reaction of the medium.

Quoting an English reaction to soilless culture we see what is a somewhat facetious quip. "Of course, it is only right and proper in a commercial age to prove all things and hold fast to that which is cheap! Of course, it is very attractive for anyone with a love of the picturesque in prophecy to foretall the time when the soil of North America—and elsewhere—having slipped away in the sea or blown away into the air, the teeming millions of that and other continents will be sustained by the produce of the Great Lakes. The charm of Lake Superior and all the rest of them covered with crops! Waving fields of maize and wheat and all kinds of vegetables, all growing in and out of the water! We, for our part, however, would rather exercise our imagination in picturing how happier and healthier we should all be if research could discover how to make the earth more fertile."

Turning more directly to the soil itself there are several factors that are being studied with profit and which have a much wider application, although not so spectacular as soilless culture.

Perhaps no phase of the soil factors has been so blandly discussed as the role of organic matter. That organic matter is built up as a bank account by the return to the soil of large quantities of manure or cover crops to be drawn on by the tree at Some future time, has not stood the test of investigation. Referring to orchard soils in particular, it has been found that the growing of green crops which were turned into the soil has resulted in a net loss of total organic matter as compared to land which was left in sod or where the trees were mulched. Tillage of orchards on rolling land has been a fallacy which has resulted in untold loss of soil and ruination of its physical characteristics which many years of intelligent husbandry will not restore.

Another result of recent studies has been the value of porosity or the movement of air and water through soil. The value of aeration in a soil has moved from its sacred place in the laboratory to a most vital position in agricultural practice. The curse of a compact soil, whether natural or brought about by injudicious culture, even in the name of a soil-building program, cannot be overestimated.

PHYSIOLOGY

The importance of physiology in the minds of horticulturists can be measured by the number who take their basic training in this field and immediately engage in investigations in which this science provides most of the tools. To even catalogue these researches would be as impossible as undesirable in this connection. It seems even more difficult to determine just what does and what does not belong in this category.

Perhaps the investigation which stimulated more thinking and additional work than any other was that of Kraus and Kraybill on the relationship of nitrogen complexes and carbohydrates on the flowering of plants. In this the plantsman found a way of thinking about plants, something he could take hold of and manipulate to his purposes.

The discovery of deficiencies of so-called rare elements, such as boron, manganese, magnesium, copper, and zinc has opened a renewed interest in nutritional studies and some important progress has already been made.

Notable in the physiological studies is the development of apparatus for the study of carbon assimilation by leaves in the field and under controlled conditions, by Heinicke and his associates. This type of apparatus affords a means of conveniently determining the photosynthetic activity of leaves which have been sprayed with various materials and of leaves which are being attacked by insects and diseases, for determining the respiration rate of plants and their products at varying temperatures and with different amounts of light. Horticultural practice has already been influenced by these findings.

From the standpoint of an orchardist, the problem of securing higher color on fruits is of serious moment, especially since the use of nitrogen as a fertiliser tends to prevent the development of sufficient red color for highest market value. Chemical and experimental evidence shows clearly that a high foliage-fruit ratio favors high color provided the synthesis of nitrogen and carbohydrates into amino and protein complexes does not deplete carbohydrates. That is a ratio of 20 leaves per apple gives a lower sugar accumulation and a poorer development of color than 75 leaves. Expressed in orchard terms, the thinning of fruit results in higher color than occurs when a tree bears an excessive crop. These matters are within the hands of the orchardist but the situation is still not satisfactory.

The discoveries and the techniques used in the discoveries represent an epochal advance in the knowledge of plants and the forces that direct plant behavior. As might be expected, there are some exaggerations of what may be expected from these discoveries; on the other hand, some of the possibilities arising from these findings would be difficult to overstate. The use of growth promoting substances in the rooting of cuttings seems to be over rated from a commercial standpoint, as does the use of Vitamin B_1 for plant growth.

REFERENCES

Dustman, R. B., and Duncan, I. J., "Effects of certain thiocyanate sprays on foliage and fruit in apples", *Plant Physics*, 1940.

Poesch, G. H., and Laurie, Alex, "The use of artificial light and reduction of the daylight period for flowering plants in the greenhouse", *Ohio Agr. Exp. Sta. Bull.*,1935.

2

Setting Horticultural Research Priorities

The future availability and prices of agricultural commodities are of interest to producers, national policy makers, traders, industry and international institutions. It is necessary to minimise and to cover risks, and to plan medium-term commodity policies. To understand what happened in the past and how it has affected the current situation, international organisations, national institutions and industry unions collect statistics on production, consumption and trade. Over the last 50 years there have been at least 30 quantitative estimates of future world food security, published regularly by the Food and Agriculture Organisation (FAO) and the US Department of Agriculture (USDA), and occasionally by bodies such as the Organisation for Economic Cooperation and Development (OECD), the International Food Policy Research Institute (IFPRI), and the International Institute for Applied Systems Analysis (IIASA).Although it is difficult to measure the precise influence of these forecasts on the actual policy debate, the fact that there are different players and models is in itself an indication of their perceived usefulness.

To look at the future of production and exchange, most agricultural forecasts are based on two types of models: trend projections (or extrapolations) and world trade models. In trend projections (e.g. of food supply and demand balances), relationships between variables lead to mathematical equations based on statistical series. The system is represented as it functioned in the past, and is used to project the future. There cannot be major changes from the past. These projections are useful for the short to medium term (one to five years).

Trend projections are temping for decision makers because they provide relatively simple answers to their questions, but they have limits. First, the description of reality provided by the model depends on the theory used. Behaviours are stylised in mathematical functions, so that the underlying hypothesis must be simple. However, as the hypothesis is not always available to the users of the model, they tend to anticipate changes that might not occur because the theories used do not correspond to reality. Second, trend projections for commodities represent only a part of economic reality. They cannot take into account the rest of the economy. If they do, through an economic model, the variables used are often exogenous. Third, in most cases, trend projections do not take into account risks – they deal strictly with the probability of things to come. At best, uncertainties are made explicit, but only in order to better isolate the statistical features of future developments.

In the agricultural sector, there are many unpredictable risks such as climate variations, wars, price fluctuations, etc. Inaccuracies in the data are another reason for the divergence between the results of a model and reality. Official national statistics do not always reflect reality, as satellite data on surfaces and yields have shown. The difficulty in representing the passage of time is another reason for the divergence between model results and reality. Changes in consumption, for example, can be much faster or slower than expected. Trend projections are therefore good for the short term, but they remain valid in the medium term only if they are constantly adapted.

Models such as IFPRI's Impact model and the FAO's World Food Model are systems of equations developed to represent how variables interact with a subsystem that has already been isolated. Estimating models based on historical trends have the advantage of transferring to the future the structure of the interrelationships among variables that were consistent in the past. Over the years, the number of issues related to the supply and demand sides of the food balance equation that are taken into account has increased significantly. Modelling approaches have become more sophisticated, larger, and much more expensive.

Foresight could be an alternative way to look at the future of agriculture and agricultural research, using a systemic approach. When socio-economic conditions evolved slowly, when technological progress was rare, farmers used their experience and knowledge to improve production and manage their environment. When those conditions started to change, farmers could no longer deal with the new context and needed help, and researchers tried to develop new techniques that they hoped would improve yields. Because of failures in the adoption of such techniques, some agronomists started to adopt a systemic approach, considering the field as a system, and then looking at interactions between the techniques. The 'systems approach' to agriculture developed from there on. Doing agricultural foresight therefore means defining and understanding the system before building scenarios or looking at innovative activities. Defining the limits of the system on which the foresight will concentrate, and defining the different elements, is often difficult due to the numerous interactions between the agricultural and other sectors.

Beyond agriculture, the notion has emerged of the national (or regional) system of innovation, which is made up of a number of actors and the links between them. Foresight offers a tool for getting the individual actors in the national innovation system to communicate, to discuss issues of long-term common interest, to coordinate their respective strategies, and perhaps even to collaborate. Foresight with scenario building, foresight models and trend projections complement each other. Once scenarios are designed, it can be useful to use a foresight model with the variables defined in the qualitative foresight, and to model the different scenarios.

TECHNOLOGY FORECASTING AND FORESIGHT

For many years, technology forecasting involved continuous monitoring of technological developments in order to identify promising future applications and assess their potential. Over the last ten years, technology foresight exercises have been conducted in Europe and Japan to identify key or critical technologies. These exercises, often government-initiated, have involved wide ranging government–industry discussions and gathering expert opinions on technology futures, leading in some cases to blueprints and government actions. In the United States critical technologies have also been identified, but through a different process.

S&T foresight looks at science in progress ('la science qui se fait')and fulfils a number of functions.It provides a means for making choices in relation to science and technology and identifying priorities. It also offers a mechanism for integrating research opportunities with economic and social needs, and thereby linking science and technology more closely with innovation, wealth creation and enhanced quality of life. It can help to stimulate communication and partnerships among researchers, research users and research funders.

Drivers of S&T Foresight

A number of reasons have been put forward to explain the growth of S&T foresight exercises, and of the various functions they are used to perform. Four major drivers have been identified.The first of these results from globalisation and growing economic competition. The growth of markets has put a premium on both innovation and knowledge-based industries and services. This in turn has given science and technology a greater importance and thus any tools that can assist in guiding investments in S&T have become important. Until the late 1980s, science operated under what might be described as the 'Vannevar Bush social contract'– governments provided funding for scientific research in the expectation that it would ultimately yield benefits in the form of wealth, health and national security, but exactly what form those benefits would take and when they would occur were rather vague. Since the 1990s government expectations of their investments in science and technology have become much more explicit and specific, so that working with stakeholders to define and execute research activities has become indispensable.

A second driver is the increasing constraints on government spending, particularly in the public sector. Governments worldwide are faced with the twin forces of a declining revenue base (associated with high economic and political costs of deficit budgeting) and the growing public demand for services such as health and education. They therefore need to be able to justify any public expenditure and to show that it is a valuable investment.

The third driver results from the enormous changes in industrial production. Command and control management has been replaced by decentralised decision making, empowerment and teamwork. Attention now is on customer-supplier relationships, and the development of strategic alliances and effective networks. The drive for controlled high performance through quality management has been extended with the new emphasis on learning and knowledge organisations.

The fourth driver has been the change in the process of knowledge production, which has become increasingly transdisciplinary and heterogeneous.The range of producers of knowledge has expanded, with an emphasis on knowledge constructed in the context of its application. In this model, there is a need for communication, networks, partnerships and collaboration in research, not only among researchers but also between researchers and research users in industry. One of the challenges facing (agricultural) researchers is to identify the potential beneficiaries or users of the results of their research, and to work with them to ensure that the results are exploited quickly and effectively to maximise the economic and social benefits.A fifth driver has been the rapid development and the growing convergence of technologies that are likely to have a revolutionary impact on economy and society in the coming decades. Frequently, however, technology foresight concentrates on identifying promising future applications and assessing their potential, rather than focusing on future research questions. In this approach, the possibility of developing new products following scientific discoveries are of central importance, as well as the conditions of emergence, development and diffusion of technological innovations.

Embedded foresight

Defining research objectives, setting priorities and allocating funds call for a long-term view. Over the years, methods have been developed for setting research priorities.Rather than continue to rely on opaque bureaucratic decision processes, S&T policy makers are starting to use S&T policy evaluation and foresight systematically to help them make choices.There is a move towards embedded foresight.

Embedded foresight refers to individual and collaborative processes through which information about relevant technological, commercial and societal developments is acquired, produced, refined or communicated within research programmes, in order to generate shared understanding of and support for research and development activities. Embedding foresight means linking it to evaluation, programme planning and priority setting. It implies the replacement of rigidly designed programmes by more flexible and learning-oriented programmes in which the priorities are continuously negotiated and revised if necessary. Such a combination enables research organisations to consider themselves not only as problem solvers, but also as the architects of new futures, and not only as researchers but as co-innovators who are involved in the implementation of future prospects.

Increased interaction between research evaluation, foresight and programme planning can enhance S&T decision-making processes. These three processes involve at least three parties – the commissioners of the exercise, internal members (i.e. evaluation or foresight group members), and external members (evaluators, stakeholders and others) – who can thus support one another. This arrangement can facilitate the formation of a common vision and ambition in which no single party – policy makers, industries, producers, community organisations or researchers – will dominate. It can create an atmosphere of dialogue and intelligent interplay between those involved, each retaining their own decision-making authority. Agreement can be reached on the division of tasks, and there can be increased cooperation in the financing, development and utilisation of science and technology. This combination can enlarge perspectives and improve policy formulation and priority setting.

For a foresight exercise embedded within an agricultural research organisation, it is recommended that the objectives are clearly stated, shared within and outside the organisation, and approved by a commissioning agency, usually a public decision maker with an interest in the work and who (co-)finances the exercise. This agency should appoint a steering committee responsible for organising and facilitating the foresight exercise, for choosing methodologies, and for compiling the final recommendations.

The foresight exercise should have a coordinator who activates potential participants, assists in the organisation of workshops and seminars, monitors ongoing efforts, and provides information. Experts from a variety of backgrounds should be invited to contribute as members of panels or brainstorming sessions and as writers of essays. These experts should be selected as individuals but represent various communities, including:

— the research organisation hosting the exercise;

— the science community at large, both to ensure that the organisation is receptive to new scientific trends and opportunities, and that new research strategies are pursued;

— consumer interest groups, since indirectly they will derive most benefit from technological change, and are potentially among the most powerful supporters of investment in agricultural research;

— the rural development community, since members of this group are confronted on a daily basis with the constraints to development and poverty alleviation, and may

— bring fresh perspectives to the key problems that must be addressed;

— the farming community - this is one of the most important groups representing the private sector, and the key reference group for most agricultural research. Farmers understand the problems and often have clear ideas about the feasibility of alternative research strategies;

— agro-industry - this is the other important group representing the private sector. In the process of agricultural development, farmers are increasingly integrated into the agro-industrial chain, through forward and backward linkages related to farm inputs, but even more to the processing and marketing of agricultural produce; the 'public' – agricultural research must relate to the problems as perceived by the public. Science journalists may contribute by identifying concerns that may be overlooked by experienced researchers; and

— national, bilateral, and multilateral investment agencies – agricultural research must be aligned with financial resource flows and agricultural investments so that research is linked to extension and development efforts.

SUPPORT OF AGRICULTURAL RESEARCH INSTITUTES

Cocoa Research Priorities

In the 1990s, the cocoa commodity chain went through turbulent times due to accelerating technical, social and economic changes. International agreements aimed at price regulation and policy harmonisation were questioned, and a new International Cocoa Agreement had to be negotiated in 1995. New relationships between public research and industry resulted in increasing vertical and horizontal integration within the sector. National research institutions faced many financial and institutional difficulties.

New scientific discoveries required new research methodologies. In this context, in January 1997 the Centre de coopération internationale en recherche agronomique pour le développement (Cirad) launched a cocoa foresight exercise to identify the research questions on which Cirad's researchers should focus, and to help them accept changes in their research priorities. The emphasis was as much on the process as on the results of the exercise.

A knowledge base, key variables and scenarios

A knowledge base covering both science and society was created, organised around four themes. Data sheets were prepared, providing information on current strengths and weaknesses, trends, possible breakthroughs and the potential for change. The data sheets were structured in such a way that would facilitate the identification of variables and trends that might influence future developments. The four themes of the knowledge base were as follows:

— *Countries.* The data sheets described the commodity chain in the producing and consuming countries, in order to assess the present situation, production trends, social needs and opportunities.

— *Actors.* The data sheets described the principal stakeholders – producers, marketing organisations, traders, industry, consumers, researchers, professional associations, international organisations, etc. – in order to assess their roles and attitudes.

— *Cross-cutting themes.* These included quality, prices, legislation, etc., in order to evaluate emerging themes for research.

— *Science.* These data sheets outlined the research results so far, current issues in various fields (genetics, agronomy, plant pathology, economics, etc.), as well as the state of scientific knowledge and the tools currently in use, in order to evaluate the situation of the research community and emerging scientific and technical opportunities.

About 20 Cirad researchers analysed the knowledge base in brainstorming and group discussions. During the exploration phase they listed and discussed potential variables. Some macro-environmental variables were characterised as 'uncontrollable' (e.g. climate, demographic trends, geopolitical shifts, economic growth and development, exchange rates). Other micro-environmental variables, such as market and industry trends, were characterised as 'controllable' by either Cirad or other actors. During the selection phase several controllable variables were identified as 'key variables' for the future. These variables were as independent as possible of each other, so that changes in one would not influence the others. The key variables for the cocoa chain were identified as: the balance between production factors (land, labour and capital), parasites and diseases, quality, consumption, and the operators of the commodity chain, especially professional organisations. Once these variables were identified, the researchers were able to prepare scenarios and to identify new questions for research.

The focus was on breakthroughs in order to develop contrasting scenarios that could provide a useful basis for the discussions. First, micro-scenarios were prepared for each of the controllable key variables that were considered to have the greatest influence on the commodity chain. Then, the compatibility of the different micro-scenarios was analysed and three broad scenarios were prepared: the continuation of current trends ('business as usual'), an 'optimistic' scenario based on the sustainable development of the chain, and a 'pessimistic' scenario that detailed a production crisis and the virtual disappearance of the chain. The objective was not to present the future accurately, but to show extreme situations in order to encourage the participants to react, discuss and decide. Mathematical simulations of the three scenarios were produced using a world food analysis security model.

The participants then analysed the roles of the many stakeholders in the cocoa chain –public and private producers, producer organisations, national policy makers, traders, industries, research organisations and intergovernmental organisations – and their relationships.

This led to the proposal of a new typology for the countries involved in the commodity chain, and to the identification of new functions in the chain.

A collective view of the possible futures for cocoa

The cocoa foresight exercise involved a sequence of collective learning cycles that took place in two processes: through the compilation of the knowledge base and bibliometric studies, and discussions. The knowledge base led to the conversion of impressions into facts. It facilitated the identification of the variables that had influenced past trends and could perhaps affect future developments, leading to a collective and shared representation of the situation. A bibliometric analysis of two internal databases was carried out to assess the strengths and weaknesses of Cirad's cocoa research programme. A retrospective analysis of publications and missions to partner countries covering ten years of activities was done. The scientists were encouraged to question the reasons for their presence in some countries and their thematic fields of research.

Discussions played an important role in the learning cycle. In a series of meetings, Cirad's researchers started to question their research activities and to wonder whether they were on the right track. They expressed some of their doubts and raised new ones. Slowly, a consensus emerged about a possible optimistic scenario for the commodity system, about changes that should be implemented, and new research activities. Outside experts were consulted on an ad hoc basis. For example, when groups of chocolate firms met, the foresight work was presented to them in order to obtain their feedback and to develop new networks. Operators in the commodity chain made useful comments that led to revisions in some hypotheses and scenarios. Individual contacts were often more constructive than group meetings because in meetings representatives of firms and policy makers would not give clear-cut opinions since they did not wish to reveal their strategies to competitors. Discussions contributed to changing opinions and to the construction of a common language.

New research priorities, partnerships and networks

After the foresight exercise, the collective view of the possible future of cocoa led to several changes within Cirad. Whereas the overall objective of the research programme had been to improve productivity, it would now focus on the sustainable development of cocoa production. This paradigm shift meant that in the future the programme would take into account economic as well as social and ecological aspects in an interdisciplinary approach. This was accepted because the researchers were aware that they had been on the wrong track.

The exercise led to the identification of threats and opportunities within the possible futures, to the definition of a desirable future, and the conditions necessary to reach it. This was a first step towards the establishment of a 'global cocoa programme'. The outcomes of the exercise were presented to stakeholders at a meeting in Paris in March 1999, and in June 2001 the Global Coordination Group on a Sustainable Cocoa Economy was officially launched. The 18 members of the group include representatives of farmers, governments of producing and consuming countries, national research centres in the producing countries, advanced and international research centres, the chocolate industry, and trade associations. Extra funding for research followed the establishment of the group. Linkages were developed between Cirad and the USDA. Since many operators in the cocoa chain were worried about future supply and consumer pressure to be environmentally friendly, they were sensitive to the added value provided by the scenarios so that they were adopted outside before internally within Cirad. All the operators decided to work in the same direction, even though each one might have had a different reason for doing so.

From a closed system into the political arena

Once the scenarios have been designed, the next challenge is to maintain the spirit and attention to outside events, and to continue to evaluate and correct ongoing activities. The foresight exercise may have ended, but the process of keeping database up to date, developing partnerships, keeping abreast of developments and, if necessary, formulating new research projects, continues. The experience of the cocoa foresight exercise shows that anticipation (prospective thought), appropriation (collective mobilisation) and action (strategic will) keep up the spirit of thinking about the future, about projects and about networks.

Many research organisations tend to function as 'closed systems',in which the goals are not imposed from the 'outside' and many staff seek to gain personal advantage, especially peer recognition. Before the foresight exercise, Cirad's cocoa programme was a closed system where researchers tended to think they alone were competent to analyse the sector's needs for innovation and to define their research priorities. They feared that the influence of stakeholder groups, especially agro-industrial firms, would limit their scientific creativity and manipulate them as researchers in a public institution. They feared being obliged to serve private interests. Evaluation and foresight pushed the 'closed system' into a 'political arena'. The differences between members of the coordination group – i.e. other stakeholders in the chain – became more apparent. The internal group became more politicised, and more concerned with its role in society. There were complex and reciprocal movements of power between the two coalitions. The foresight became the basis for a new type of discussion. Foresight and evaluation are indispensable instruments for

opponent on this issue, Albert Einstein, he explains how the discussion of the epistemological problems has shown that it is not possible to comprehend the evidence of quantum phenomena, obtained under different experimental conditions, within a single picture. The evidence from different experiments must be regarded as complementary in the sense that only the totality of the phenomena exhausts the possible information about the atomic objects. The complementary pictures of quantum phenomena are inseparable from the observational situation – it is impossible to separate the behavior of atomic objects from the interaction with the measuring instruments that serve to define the conditions under which the phenomena appear. And Bohr advocated the application of the very word *phenomenon* exclusively to refer to the observations obtained under specific circumstances, including an account of the whole experimental arrangement.

This meant that Bohr had to dismiss the "realistic" presumption of a distinction between mutually exclusive subjects and objects entailed in the conventional conception of objectivity. Subsequently Bohr's "Copenhagen interpretation" of quantum theory was derided as "subjectivist." But his philosophical analysis of the observational situation in atomic physics still stands, and it suggests a new conception of objectivity. According to Bohr, the inclusion of an account of the experimental arrangement and the results of observation in the language of everyday life and classical physics was the only way to provide an *unambiguous description* of the experiences in atomic physics, where the phenomena far transcend the classical physical experiences. And to Bohr, objectivity meant simply the unambiguous description of experiences. Furthermore he suggested that the situation in physical science was analogous to the situation as regards analysis and synthesis of experience in many other fields of human knowledge and interest, such as psychology and life sciences.

The conception of research as a cognitive system does not presuppose a distinction between subject and object. It focuses on the systemic connections of research. Below, a model of the experiential situation in research is outlined that makes the role of science as both actor and observer more explicit. This self-reflexive model forms the basis for understanding how the cognitive experience that is gained in research is "objectified."

Self-reflexive Circle of Learning

Looking at research as a cognitive system, we can picture the learning process of systemic research as a self-reflexive circle – a cyclic cognitive process including the representation of oneself as another. This model is in analogy with human self-conscious learning based on the ability to take a mental step out and look upon oneself and one's actions from outside, and use this outside view in later action. Selfreflection starts from the viewpoint, or stance, of the "actor" – the first order involved viewpoint of a cognitive system (with the capacity for self-reflection) – then it moves to a second order detached viewpoint where the "observer" views the system from outside. And the observations made from this outside point of view can take effect upon returning to the first order viewpoint of the system.

The distinction between inside and outside viewpoints has been widely, although not consistently, used in anthropology and other fields under the names emic (inside) and etic (outside). The terms actor and observer, or the simple terms inside and outside viewpoint, for several reasons. First of all, they are easily understood and interpreted in a common way, as opposed to emic/etic, and they can be linked to the guiding metaphor of self-reflection in a straightforward way. It might be tempting to use the terms subjective and objective, but there is a load of connotations in those terms that are better avoided in this context for the sake of conceptual precision.

In our understanding, doing systemic research involves joggling these two different points of view. When a science recognizes itself as systemic, that is, as influencing its own subject area, this leads to reflection on how it interacts with the world. And the model in figure 2 can support this growth of reflexivity. Moving from an inside to an outside viewpoint entails that an overall distinction between the system and its environment needs to be made – the system has to be identified as an object of observation. This first movement also involves the determination, or at least presumption, of certain goals and values upon which the choices and delimitations that need to be made in planning and initiating research, can be made.

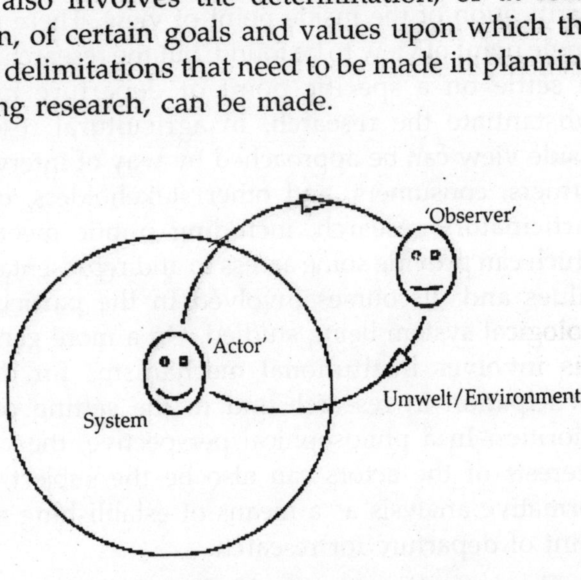

Figure 2: The self-reflexive circle of learning in systemic research, moving from an inside actor viewpoint, or stance, to an outside observer viewpoint, and back.

The ensuing observations are thus based on these value-laden choices. These choices, together with the initial choice of system to be studied, determine the relevance of the research. The questions of for whom the research is done and what intentions guide the research are determined here. In other words, if the observations are to be part of a learning process for the observed system, or for some specific social system, the choices made need to reflect the values and goals in that system, because these choices can be decisive when the observations are later used in the development of the system.

Sometimes the researchers are, in a sense, already within the system that they are researching, because they share the worldview, values, and goals present in the system in a way sufficient to initiate the research. But often there is a need for a reflexive determination and clarification of the inside point of view. There may be no single point of view to be found, but the research unit needs to settle on a specific point of departure in order to substantiate the research. In agricultural research, the inside view can be approached by way of interviews with farmers, consumers, and other stakeholders, or through participatory research, including public meetings, etc., which can provide some access to and representation of the values and discourses involved in the particular socio-ecological system being studied. On a more general scale, this involves institutional mechanisms for democratic participation in research and in the setting of research priorities. In a philosophical perspective, the values and interests of the actors can also be the subject of critical normative analysis as a means of establishing a coherent point of departure for research.

The second movement, back to the inside viewpoint, may take direct or indirect paths, depending on the relation between the system and the research unit. In research as a cognitive learning process, a direct path is taken in the form of actions that transform the system. The indirect path goes by way of communication. As well recognized, the observer stance forms the basis for scientific communication. But the communicational learning process in science is dependent on the quality and adequacy of the connection to the underlying cognitive process. In other words, it is dependent on how the cognitive experience is "objectified."

Taking an outside viewpoint is the hallmark of science and also an indispensable part of systemic research. The outside viewpoint first of all detaches the interests and goals of the observer from the dynamics of the system. Moving from an inside to an outside view in systemic research, approaching the position of the "objective" observer, therefore allows for a distinction between dependent and independent dynamics in the learning process. The observed system dynamics are henceforth uninfluenced by the observer's intentions, and this allows the observer to learn about the independent dynamics of the system. But, given the self-reflexive circle of learning, the outside or "objective" stance always rests on a specific inside point of departure – it is not the privileged, detached, value-free, Archimedean point of observation that is entailed in the conventional criterion, or ideal, of objectivity. Therefore, the independence of the observed dynamics of the system can only be a conditional independence, an independence that is conditional on the choices made in order to take an outside viewpoint.

Based on the understanding of research as the learning process of a cognitive system that has been presented above, the model of self-reflexive learning in figure 2 applies to all research. All research depends both on cognitive experience and the transformation of those experiences into scientific communication. In other words, all research starts with a complex system that embodies values, namely the research unit and its world. In laboratory research, the research unit acts in the construction of equipment and experiments. In the observational and historical sciences the research unit acts in the choice and development of tools of observation and the construction of coherent narratives. Research that considers itself only a detached observer of the world without also being an actor is thus blind to parts of its own function. On the other hand, research that operates only as an involved actor fails to be scientific.

In a critical perspective, self-reflection allows research to make an "objective" description of the specific value-laden point of departure, which can be communicated as a part of the context of the observations. However, it is by no means an obvious or easy task to identify this context.

"Reflexive objectivity," Relevance, and Cognitive Context

The purpose of peer criticism in the scientific community (as opposed to the general discourse of society) is to subject the knowledge of the individual researcher or research group to the perspectives of other researchers, and thereby root out errors due to the influence of the individual perspectives and peculiarities. And, as stated above, one of the distinct features of scientific learning is the open, public nature of scientific communication. As generally recognized, however, peer criticism is only in principle open to the scientific community in general due to the specialization of science. In practice, peer criticism is therefore to some degree conditional on special knowledge and restricted to disciplines or sub-disciplines. However, if the scientific communication is to be even *in principle* open, it has to include sufficient context to be unambiguous, as stressed by Bohr. And, insofar as the value-laden context of scientific inquiry is an important aspect of the knowledge production of science, this aspect has also to be included in the scientific communication in order to maintain the ideal of openness. Even though this seems to be in conflict with the conventional scientific ideal of objectivity, which implies independence of contextual values.

With reference to Luhmann and Bohr, the experience of the research unit becomes "objectified" by being transformed into scientific communication. In this transformation process, the researcher approaches the stance of an "ideal" observer so that this "objective experience," in principle, can be shared with and criticized by any member of the scientific communicational community. The means of "objectifying" experience are the conventional scientific methods of documentation and control of observations and experiments, etc. The demarcation of science as a special learning process, it suggest, is the very ability to take an "objective" stance. Not the conventional, unreflected objectivity, which excludes the intentional and value-laden aspects of science, but a *reflexive objectivity* that includes these aspects and exposes their role.

The characteristic of reflexive objectivity is that the false assertion of value-freedom is considered less objective than the admittance of the contextual values and interests involved in research, because this context can be decisive for the learning process by way of the choices and delimitations made. The value-laden starting point should therefore be well described and clear, and included as a necessary context in the subsequent communication of "objective experiences."

Where the conventional criterion of objectivity seems to stand in opposition to the criterion of relevance, reflexive objectivity is interrelated with relevance through a common focus on context: Relevance concerns how the context of research is formed, and reflexive objectivity concerns how the context of research is exposed. In effect, reflexive objectivity demarcates the outer boundary of what is good science, while relevance determines the research that is actually to be performed.

In the perspective of science as a self-reflexive learning process and the criterion of reflexive objectivity, the answers that science gives depend on the *cognitive context*. The cognitive context can be divided into at least three levels (Table 1). The *societal context* is the group or social system that the term "relevance" refers to, that is, those or that which the research is supposed to be relevant for (e.g., society, farmers, the third world, the market, the employer, the sponsor, or the scientific community). The *intentional context* consists of the goals and values that guide research, including what is taken as relevant problematic situations. The *observational context* includes the actual distinctions, delimitations, models, and technological constructions that are employed in research. That is, the means that are needed to perform research and obtain observations. When Bohr defined "phenomenon" as referring to the observations obtained under specific circumstances, he pointed to the observational context. The term "levels" as an indication of the structure of conditional independence of the observations and types of context.

Table 1: Three levels of cognitive context

Level of context	*Description*	*Examples*
Societal	The social system that the term "relevance" refers to	Society, particular social groups, the sponsor, science
Intentional	The values and goals that guide research	Visions, aims, problems, etc.
Observational	Conceptual and technological tools that are needed to perform specific research	Cognitive schemata, concepts, models, instruments, labs, etc.

The observations are influenced by the observational context, while they are independent of the other levels of context given the observational. The observational context is influenced by the intentional context and it is independent of the social context given the intentional. And finally the intentional context is influenced by the societal. For instance, when considering research as a tool for developing organic agriculture, the societal context is first of all the organic movement. The organic movement has a set of rather explicit visions, values, and goals that frame the intentional context. And this, at least to some extent, determines the observational context. On the other hand, some of the organic values are relevant in a wider societal context. And the research that is actually performed can very well be relevant to other farming traditions with other goals and values.

Usually, peer criticism is primarily directed towards the relation between the research results and the observational context. However, given the above structuring of the context of research, there is also a role for critique of how the observational context is related to the intentional context. For example, is the intentional context made clear? Do the chosen approaches, methods, and concepts address the problematic situation or the expressed goals and values in a fruitful way? The meaning of the evaluative and normative concepts that are used in science, such as soil quality, animal welfare, and sustainability, is a well-known problematic issue. The exposure of the values embedded in the usage of such concepts is an important aspect of reflexive objectivity. There is furthermore a role for critique of how the intentional context is related to the societal context. For example, do the stated goals, values, and problems reflect the concerns of the social systems that the research is to be relevant for? Is the funding of the research made clear or hidden? In addition to peer criticism, which concerns the exposure of the cognitive context and the coherence of context and results in relation to reflexive objectivity, there is a role for public criticism, which directly concerns the contextual values: Is the exposed context found to be relevant?

Peer criticism is often conceived as the peer review of papers as part of the publishing process. But there is an important role for criticism in other parts of the research process. In particular, new research projects are determined through a process that involves politics and funding. In this process, the relevance of research can be improved by way of exposure and communication of the suggested cognitive context. The task of making clear and criticizing the intentional context and the relation to the societal context involves value inquiry. This is not an easy or straightforward task, but it is a substantial and important element in a systemic research methodology.

It is argued above that, given a particular well-described observational context, the values and intentions that led to this particular setup will not influence the results. If this is so, one might ask why we should care about the intentional context. Exposing the intentional context is, however, an important background for seeing the particular observational context, and the "objective" knowledge that is gained, in a larger perspective. It will facilitate the coordination and cooperation with other research that has a compatible intentional context, and allow groups with certain values to identify research that is compatible with those values and use the research with confidence in the relevance of the results. Similarly, exposing the societal context (e.g., for whom is the research done, funding, etc.) will allow for a normative and political discourse on the intentional context and thereby provide the structure that enables the functioning of science as a genuine learning process for society.

Some might argue that values should, in fact, *not* play a role in the method of science, because there is only one objective scientific approach to any given problem, or because science is not a problem-driven but an explorative, descriptive enterprise. These arguments build on a very narrow conception of science that disregards the role of science as a tool for action and development. There is always a (reflected or un-reflected) choice of subject matter or research issue in science. If science is seen as problem-driven and a tool for action, this choice depends on what is taken as problematic, and this involves goals, interests, and values. And even the enterprise of "pure science" entails certain values that may be questioned in relation to the role of science in society.

Values also play a role in the method or approach employed in addressing a specific problematic issue. The delimitations and reductions employed and the choice of methods are intertwined with value-laden issues. As a fictional example, let us imagine the general problem of fungus diseases in wheat crops (a problematic issue that already presumes certain values and perspectives). There are many ways to do research in regard to this problem:

— you can develop pesticides that are (more or less selectively) effective against fungi

— you can search for and breed new sorts of wheat that are more resistant towards fungi, or use genetic engineering to develop wheat that is toxic towards fungi - you can investigate different agricultural practices (in terms of levels of fertilisers, etc.) that make the wheat more resistant towards fungi

— you can investigate different agricultural production systems where the wheat is less prone to get fungus diseases in the first place (through crop diversity, landscape diversity, etc.)

— you can investigate the use of other crops than wheat

— or you may even investigate other land uses than crop production, depending on the interests and goals of society and stakeholders.

Such choices are made in research initiation, and the choice of approach or method indicates something about the intentional context that is (knowingly or unknowingly) employed. The relevance of the research depends on whether these choices correspond to the values present in agriculture and society, or whatever specific intentional and societal context "relevance" refers to.

In short, relevance is concerned with research as a tool for development with respect to certain groups with certain goals, interests, and values, while reflexive objectivity is concerned with making clear the context of the learning process. The criterion of reflexive objectivity suggests that research should investigate and describe its own societal, intentional, and observational context and work explicitly with the goals and values involved, in order to facilitate peer criticism and the use and critique by different users and stakeholders. In particular, research should describe the choices made in research initiation, the delimitations and constraints these choices place on the results, and the areas of ignorance that this implies, as an essential context of the results produced. Such changes in the norms of science cannot be implemented by single researchers or research groups. Their successful implementation involves all the different institutional structures of science, such as the organizational structures of research, the media of publication, the structures of research policy and funding, the educational institutions, etc.

A Common Framework for Transdisciplinary Systemic Research

A unitary view of science is a learning process. This view does not presuppose different kinds of science, such as the established division between natural, social and human science. A second important perspective for a systemic research methodology is therefore the relation between the different and often quite separate disciplines in the highly fragmented science of today. In order for science to function as a common learning process, there is a need for a common framework that can serve as a basis for doing transdisciplinary systemic research.

The "boundary work" of science, the actual demarcations of science from non-science that science employs in the pursuit of its professional goals, has implications for the conditions for doing transdisciplinary systemic research, which needs to work across the established boundaries of science. Cross-disciplinary research is often hampered by the idea that reductive science is more objective, and hence more scientific, than the less reductive sciences of complex subject areas – and by the opposite idea that reductive science is necessarily reductionistic. This has led to an unfruitful opposition between"reductionist" and "holistic" science in connection with, for instance, agricultural and ecological research.

From the holist (or anti-reductionist) point of view, analytic, reductive methods are bad science because they do not capture the connectedness of complex reality. And reductionist science is (in part) to blame for the present agricultural and environmental problems. From the reductionist point of view, analytic, reductive methods ensure the quality of science, and other methods are, therefore, not scientific. Hence the term "holistic" seems to promise more of science and cognition than can be fulfilled. Since reduction is a powerful approach in science that can contribute significantly to the learning process, the term "reductionist" should be used only where a science is unaware of the consequences of reduction or denies that there are any such consequences. Given an evolutionary metaphysics that asserts the reality of emergence, reductionist science is to be regarded as bad science. Given even the hypothesis that there are emergent phenomena, a science that does not investigate what may be left out in reduction, does not meet the quality criterion of reflexive objectivity.

The condition for avoiding reductionism is a reflexive awareness of the observational context, including the analytic methods. The conditions for being "wholeness-oriented" are, in addition, a reflexive awareness of the role of specific disciplines and research activities in relation to the larger wholes or systems that are found to entail problematic issues (e.g., environmental problems), and an awareness of the role of science in society. In the perspective of organic agriculture, which takes the connectedness of the world as a basic presumption and the independence of things as something special, there is a particular emphasis on the problems related to science being fragmented by disciplinary specialization without enough awareness of the need for interdisciplinary and transdisciplinary work and for methodological and organizational structures that can facilitate such work.

Our conception of science as systemic and the promotion of wholeness-orientation is largely in congruence with Michael Gibbons *et al.*'s description of how the production of knowledge (scientific learning) is advancing into a new phase. From the conventional natural science-like knowledge production has evolved a new mode, which continues to exist alongside the old mode. Mode 1 is disciplinebased and carries a distinction between basic and applied sciences, with a linear operational relation between a fundamental theoretical core of knowledge and its practical use in applied sciences. By contrast, Mode 2 knowledge production is transdisciplinary – it is characterized by a constant flow back and forth between the basic and the applied, between the theoretical and the practical. In Mode 2, more emphasis is placed on contextualized results and less on the search for fundamental principles, and the processes of discovery are integrated with those of fabrication. Where the problems of science in Mode 1 are set largely by the academic interests of a scientific community, Mode 2 is fuelled by practical problems in specific contexts, involving heterogeneous groups of participants.

In the following, the unitary view of science as a learning process is linked to a framework that can assist transdisciplinary work. The framework locates different kinds of science in accordance with the kind of motivation or interest behind the research and their main methodological characteristics.

Classifications of Science-Based on Motivation and Methodology

The notions of systemic science, wholeness-oriented

research, and Mode 2 knowledge production seem to capture essential features of an important shift in the conception of what science is as well as in the general methodology and structure of science. The shift may also involve changes in how science is divided into different kinds. This shift is related to the involvement of science in new complex areas, such as the socio-ecological problems of society and environment. Here, different disciplines such as ethics, sociology, ecology, and chemistry ought to be in close cooperation. This is not often the case.

Donald Stokes has recently, in a discussion of the basic/ applied distinction, suggested a distinction between four kinds of science that is based on two dimensions of *motivating factors* for research: the considerations of use and the quest for fundamental understanding of the causes of phenomena. Stokes's critique of the basic/applied distinction is well taken, and his identification of two different dimensions in this distinction seems to address a real problem with it. There are, however, some difficulties in his fourfold classification. The dimension of being motivated by a quest for fundamental understanding seems to conflate two different *methodological* dimensions of science, the degree of reduction and the degree of observational detachment.

In an analysis of the relation between knowledge and interest, Jürgen Habermas distinguishes between three kinds of science. This is a distinction that might resemble the traditional distinction between natural, human and social sciences, but it is one that is not based on the character of the subject area. Habermas's distinction refers both to methodology and to motivations. The three kinds of science are 1) the empirical-analytical sciences with a cognitive interest in *technical* control over objectified processes, 2) the historical- hermeneutic sciences with a *practical* cognitive interest directed towards mutual understanding and meaning, and 3) the systematic sciences of social action with an *emancipatory* cognitive interest.

These latter sciences, including economy, sociology, and political science, not only produce nomological knowledge of causes and laws — they are also critical social sciences by way of the method of self-reflection in society, because self-reflection "releases the subject from dependence on hypostatized powers". The very awareness of the causal mechanisms allows for the liberation from these mechanisms. Critically oriented sciences share this emancipatory cognitive interest and the method of selfreflection with philosophy.

A framework for the cooperative understanding of different kinds of science

If we look at, for instance, laboratory research, field experiments, on-farm research, and sociological research in agriculture, they are different with respect to the reduction and abstraction of the subject matter, and subsequently in the conditions for experimentation and control, and thereby for replications or reproductions of phenomena. They are also different with respect to the need for ethical considerations.

As stated above, there has been a more or less tacit conception of science, where systemic approaches that include, for example, the human and social parts of the agricultural systems into their research world, are perceived as less scientific than conventional, analytical approaches, which have reduced their research worlds to exclude those aspects of reality. There *are* obvious differences between agricultural systems research and analytical chemistry, for example, but sciences in some subject areas are not necessarily more scientific than sciences in other areas — there is no necessary difference in the potential for doing good science in more and less reduced subject areas. And in a learning perspective, the possibility for doing good science does not increase with the amount and "solidity" of existing knowledge and theories.

The important methodological dimensions of such a common framework are, first, the dimension of more or less reduced "research worlds" and, second, the dimension of detached observational knowledge versus involved experimental action. The latter dimension refers to the two stances of the learning circle, the detached observer stance, where general "objective" knowledge can be generated, and the involved actor stance, where decisions and actions actually take place in specific contexts. With respect to the first dimension, the reduction of subject matter in science involves a construction of a simplified research world by way of abstraction and metaphorical construal ("gedankenexperimente") and mostly also by way of physical delimitation, manipulation, and control (classical experimental science). The conception of reduced research worlds is in line with Joseph Rouse, who rethinks the role of laboratories in our understanding of science.

> Fundamentally, a laboratory is a locus for the construction of phenomenal microworlds. Systems of objects are constructed under known circumstances and isolated from other influences so that they can be manipulated and kept track of. They constitute attempts to circumvent the chaotic complexity . . . by constructing artificially simplified 'worlds'. In these microworlds there exists only a limited variety of objects, whose provenance is known and whose forms of interaction are strictly constrained.

The common framework of science that emerges from the above considerations is illustrated in figure 3. The two methodological dimensions span the framework: vertically the more or less reductive sciences, with the very reduced research worlds at the top, and horizontally the involved and the detached stance. The triangular form represents the effect of reducing complexity and working with restricted research worlds. The more detached, observational kinds of research can be classificatory or descriptive and historical.

And the more involved, experiential kinds of research can be experimental or developmental. In highly reduced research worlds, the classificatory and experimental sciences are so closely connected that they cannot be separated and hardly distinguished. In less reduced worlds, on the other hand, the descriptive and developmental sciences are very different and hardly seen as connected.

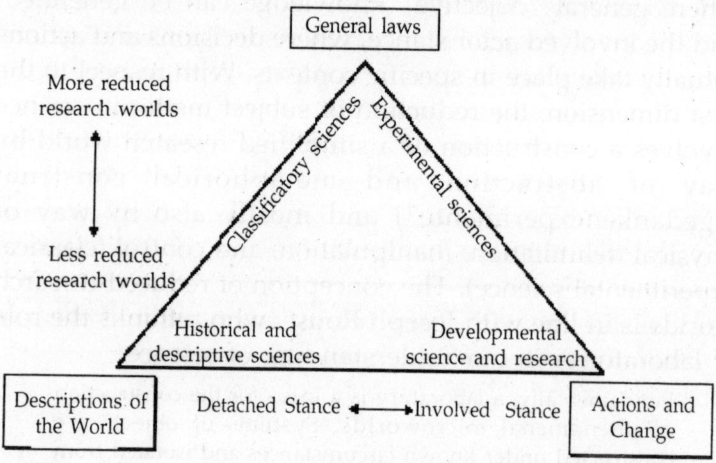

Figure 3: An illustration of the relation between different types of empirical science.

The two aspects of science, the empirical and the normative, are shown as two faces of a pyramidal structure. Each face is spanned by the two methodological dimensions. In this framework, some examples of different disciplines and types of research are indicated that are of relevance to systemic research in agriculture. With respect to other subject areas, other examples would be appropriate. Even though the examples are placed on one of the two faces, none are purely empirical or purely normative. Every actual field of science has both normative and empirical aspects even though the focus may be entirely on one or the other. In particular, among the highly involved systemic or developmental sciences, action research has substantial normative aspects and political theory has substantial empirical aspects.

Among the more reductive normative sciences are the sciences of good thinking and good inquiry. Moving upwards in the empirical triangle can be aligned with seeking a reductive, causal explanation of a phenomenon, while moving towards the detached can be aligned with seeking a classification or a communicable description of a phenomena. "General laws," such as the natural laws of physics, can be taken as the extreme expression of the method of reduction in science, providing an ideal causal explanation. They represent a general knowledge that is valid everywhere, but it is only valid given the constraining presumptions in the reduced research world. If you let an egg fall from the leaning tower of Pisa, it will fall in accordance with the laws of gravity. If you let a feather fall, more complex laws interfere. If you let a swallow fall, the laws of gravity play only a minor role in explaining what happens.

In more complex worlds, the epistemological limitation with regard to using causal knowledge from reduced worlds is that some aspects are neglected because of the reduction. Some examples of neglected aspects are the crop rotation effects when doing research on single fields; the effects on farm economics when doing research on cropping systems; the management factor when studying farm dynamics in experimental farming systems or in the assessment of animal welfare; or the role of motivation in behavioural studies.

Conversely, the methodological and technical limits of doing research in more complex research worlds are due to the complexity and variety of the research world and the diversity and individuality of the entities. For example, it is very difficult to make a general predictive model of a complex world – in particular if the world involves selfreflexive actors, such as persons and their social systems, that learn, adapt, and change course in accordance with the knowledge they gain from that model.

There are also ethical limits to research in complex worlds, connected to the presence of human and non-human actors in the research world that the research system has a moral responsibility to take into consideration and, in the case of persons, involve in a moral discourse. In this perspective, reduction entails that the ethical questions are externalised–they become part of the external communications of science and the actual research can be done without ethical considerations. In a wider systemic perspective, however, where science is seen as a part of society and nature, and with reference to the criterion of relevance or given a theory of moral responsibility for actions, there can be decisive ethical concerns connected with the potential consequences of reductive science.

REFERENCES

Alroe, H. F., "Science as systems learning. Some reflections on the cognitive and communicational aspects of science", *Cybernetics and Human Knowing,* 2000.

Bawden, R. J., "Systems approaches to agricultural development: The Hawkesbury experience", *Agricultural Systems,* 1992.

Checkland, P. B., *Systems thinking, systems practice,* John Wiley & Sons, 1981.

Kuhn, T .S., *The structure of scientific .revolutions,* 3. Ed.Chicago: The University of Chicago Press, 1996.

4

Partnerships in Horticultural Research

Horticultural research is conducted primarily in the public sector, with research at private institutions playing a relatively minor role. As a result, research gaps naturally emerge between the basic research generated by public institutions and the research needs of industry. One approach for reducing this gap is to form public-private research partnerships that harness the complementary research and academic expertise of universities with the commercialisation and marketing expertise found in industry.

Such partnerships are proliferating, especially between universities and large life-sciences companies. Unfortunately, there are few concrete examples of such partnerships in agricultural biotechnology for the horticulture industry. The challenge is to adapt models of these partnerships to the research needs and structure of the horticulture industry, which produces crops such as fruits and vegetables, nuts, and nursery and ornamental crops. The traditional research paradigm posits a one-way flow from basic science conducted in public institutions to applied research and commercialisation undertaken largely by private industry. This characterisation does not accurately portray current trends in research and development (R&D).

Increasingly, public universities and private firms engage in joint research and establish interactive relationships. Several factors have contributed to this trend, including recent legislation, the restructuring of many of the larger life-sciences firms (such as Monsanto and Syngenta) and an alignment of private and public incentives to pursue long-term R&D efforts.

The potential benefits from universityindustry partnerships in the field of agricultural biotechnology are obvious. Scientific and practical knowledge can complement each other, leading to more rapid and far-reaching innovation. Universities need funding for their researchers, as well as intellectual property held by private companies and access to modern, commercially developed enabling technologies to ensure a first-rate graduate education for students. For its part, industry is interested in accessing new research and innovation, developing new products and hiring highly trained graduate students.

However, obstacles to the formation of successful agreements are significant. Both parties in a research partnership face serious risks. These risks are rooted in the conflict between a university's academic objectives and the private firm's corporate incentives. One critical risk is the potential co-opting of the academic research agenda by private interests. University researchers risk the loss of academic freedom and integrity while industry risks the loss of investment capital, privacy and proprietary information.

Differences between the university's educational objectives and corporate goals, as well as differences in the cultures, institutional incentives and time frames, can lead to a clash of cultures and values. Intellectual property (IP) rights issues are also a frequent source of contention. Given these risks, both parties need to enter into carefully structured research agreements.

STRUCTURING AGREEMENTS

Most work examining research partnerships focuses either broadly, on such issues as the source of research funding, basic provisions of these agreements and associated problems and consequences, or narrowly, on specific aspects of a particular type of agreement. Although this literature is useful, it does not effectively address how to structure these public-private research partnerships. In response to this need, we have constructed templates based on the three stages of any universityindustry research partnerships, which provide a framework for characterising their "front-end" and "back-end" options.

University-industry research partnerships come in many forms. They may be targeted, with private firms designating specific research agendas, or they may be nontargeted. Research projects may have short or longer time horizons. Universities may enter agreements with a single private company or with groups of firms sharing a common interest. Collaborations may cover a single research project or be "megaagreements" covering a large range of interactions. Because of the inherent uncertainty in the research process, research partnerships can be structured in terms of ex ante decisions on the options embedded in the three stages of any agreement. These embedded options are specific decision points, such as determining which partner will control the research agenda. Universities can define policies on this option ex ante, before potential partners are approached.

Stage I: Setting the bargaining space. To start, potential research partners consider possible collaborations and associated tradeoffs. The vital aspect of this stage is determining exactly how partners will be identified and selected.

Although deliberately seeking out partners rather than waiting to be approached with a proposal requires more effort upfront, it can substantially broaden the set of choices. For example, the public partner could elicit competitive bids from multiple private partners rather than just accepting or rejecting a single proposal.

Stage II: Negotiating the agreement. Next, the agreement is negotiated and may or may not involve a formal contract. Front-end options determine the nature and scope of the research activities that the partnership will undertake, while back-end options determine how any benefits generated by the partnership will be distributed and how knowledge assets such as patents and commercial products are disseminated.

Decisions in the front-end include specifying the research agenda, asset contributions, governance structures structures and scale of operations. Back-end options include designating patentfiling responsibility, property and licensing rights, royalty rates and how research results will be disseminated.

Stage III: Reviewing and renewing the partnership. Finally, the outcome of the partnership is assessed, with an eye toward whether to renew the agreement. Currently, there is no standard approach for formal review of large- or small-scale agreements. To assess whether a research partnership was successful or not, interested parties must rely on the informal reviews and vague impressions of both partners along with more tangible outcomes, such as the number of patents generated by the research. A key policy challenge is the development of concrete indicators or measures of productivity for public-private research partnerships.

Templates for Partnerships

Based on these stages of forming agreements, we have designated four groups of templates. Strategic partnerships involve comprehensive, multiyear commitments between a university, or an academic department in a university, and a large company, with both partners dedicating significant assets. Formal procedures for determining research agendas and control of back-end assets are specified.

Research unit/center partnerships usually also involve the dedication of significant resources. Instead of involving existing academic departments, however, these research units are set up separately, allowing more distance between the partnership and the academic community at the university.

Such partnerships may be linked to a single company, commodity group or companies that provide some or all of the financial resources for the research center. For example, the Seed Biotechnology Center at UC Davis is a partnership between the College of Agricultural and Environmental Sciences and the California seed industry. The College provides research space and faculty time, while the industry funds additional research and programmatic personnel. Specific research projects are funded through diverse grants and contracts with both public agencies and private sources.

Sponsored projects are small to large commitments with a specific research agenda designated at the outset. As with strategic partnerships, either partner may approach the other, but instead of defining a governing structure for selecting research directions, specific research projects of particular duration and budget are proposed. Depending on the nature of the bargaining space (e.g., private partner proposes project versus the university approaches private partner with research needs), the university's options on the front-end can be more restricted. Sponsored projects may act as testing grounds for relationships and serve as precursors for more far-reaching strategic partnerships.

Horticultural Industry and Research

The horticultural research industry is composed primarily of small to medium enterprises with small markets for individual products. Because of their relatively smaller size, these firms are able to rapidly apply new knowledge and technology. However, when it comes to genetically engineered crops, the smaller firms generally do not have the assets to develop new products.

Research funds in horticulture come mainly from the public sector. The reluctance of major biotechnology R&D companies to dedicate funds to horticultural research is, in part, because technological advances in horticulture are not viewed as "low-hanging fruit." The commercial value is not nearly as attractive as for annual agronomic crops grown on large acreages. In addition, consumer acceptance of genetically modified foods is considered a major obstacle to the adoption and commercialisation of agricultural biotechnology. Current biotech research focuses on reducing the environmental impacts of horticultural production, food safety, product quality and newproduct development.

Public-private research partnerships could greatly benefit the horticulture industry, and domestic and international research partnerships in horticulture are considered especially important for developing economies. Dixon notes that successful entrepreneurs in horticulture maintain a continuous dialogue with scientists; partnerships are one approach for guaranteeing this dialogue. Dixon also notes that linkages between research and industry have improved "where levy funding systems have been established to support scientific endeavors." In other words, more formal financial arrangements between partners are likely to yield the best exchange.

Strategies for Horticulture R&D

The most relevant partnership model for the horticulture industry is that of less formal, single or multipleproject partnerships. In pursuing these partnerships, the implications of all three stages of the partnership should be considered ex ante.

In Stage I, private horticulture institutions seek to align research incentives and form consortia of small and medium firms with parallel research interests to concentrate intellectual and financial resources. These consortia are organised by crop or pest type to facilitate networking, identify key researchers at public institutions and propose specific research projects.

The university should accept or reject these proposals based on the research synergy and embedded options. Although all universities share a common set of core principles that guide their decisions, different institutions emphasise different objectives; the private partner should consider the university's research climate when considering research partners.

In Stage II, the private partner considers the type of research to pursue in the partnership. Given the nature of research objectives at universities, the horticulture industry partner should propose research projects that are more basic, have longer time frames and are not adequately addressed by current private research efforts. These partnerships are more likely to be successfully negotiated if the industry partner understands, ex ante, the need for faculty and academic freedom. On the backend, university guidelines and policy usually constrain its researchers to specific conditions for patenting research, and licensing and disseminating results. Although there is some variation, these constraints are fairly common at universities. Stage III is best accomplished if specific goals or benchmarks are incorporated into the initial agreement. This gives both sides criteria to judge whether the partnership is achieving its goals and justifies renewal.

Consortia Benefits and Risks

Both partners should establish links so that industry can

more effectively utilise public research and universities can secure access to research funding and cutting-edge enabling technologies. These collaborations can serve as stepping-stones to more formal, long-term agreements. Alternatively, once initial consortia-university research partnerships are established, more enterprising members of the consortia can capitalise on the research contacts and pursue firmspecific, applied-research partnerships.

The primary obstacle to forming research partnerships is high transaction costs. The process of identifying appropriate researchers as potential partners can involve significant search costs. And once potential partners have been selected, the time and effort involved in negotiating a research agreement, especially given the differing objectives of public versus private institutions, can be substantial. The consortium approach is a strategy for sharing these costs. If the consortia are not well structured, however, reduced external transaction costs may be replaced by higher internal costs of organising and running the consortia. Inequitable benefitssharing within a consortium may also be a source of conflict. And although this approach is intended to serve the needs of medium- to smaller-sized firms, the smallest enterprises may still be excluded.

FUNDING HORTICULTURAL RESEARCH

Agriculture the world over has changed from a resource-based to a science-based system. Technical change has accounted for half to more than four-fifth of agricultural output growth since the sixties in developed countries. Technology-led growth requires investments in technology generation, extension, education, physical and institutional infrastructure.

A large and diverse country like India cannot depend on borrowing technologies from other countries. Agricultural production technologies are generally less transferable than others. Research output is often a public good. Market failures are common and appropriability of benefits is restrictied, thus limiting incentives for private initiative. Also private sector is mainly driven by short-term profit considerations, longer term or sustainability concerns are rarely accorded priority. Therefore, even in highly developed countries, strong public research systems have to be supported.

How Much to Invest

Use of conventional investment criteria to determine appropirate research investment levels is fraught with many conceptual and operational difficulties. Rates of return studies suggest substantial underinvestment across the board. These typically show 40+ percent returns even in countries where research investment intensities are five times higher as compared to developing countries.

For India, less than 0.5 percent of agricultural GDP is spent on research. At Independence, this was less than 0.1 percent, rising to 0.12 percent in 1960, and about 0.2 percent towards the end of sixties. After a spurt through the seventies, it has hovered around 0.45-0.50 percent since early eighties. This implies that research investments have barely kept pace with growth of agriculture; there has been no special accent on jacking up research investment intensity over the last 15 years.

It has been estimated that in developing countries as a group, research .investments account for about 6-8 percent of total public spendings on agriculture. Since the centre accounts for nearly 60 percent of total public spending on agricultural research, signals on low priority filter down to states also. Analysis of trends in developed countries which spend close to 2 percent or more showed that between 1965 and 1985, there was positive growth in real expenditures per scientist, implying distinct improvement in research support to scientists. In 130 developing countries, on the other hand, there was negative growth in real expenditures per scientist, suggestive of horizontal rather than vertical expansion of the research system. Accordingly, there has been a deterioration in availability of operating funds per scientist. This has been the case for India as well.

Over the Ninth and subsequent plans, these trends must be corrected and a target level of 7-8 percent of total agricultural outlay for agricultural research and education must be established. With concomitant growth in private sector spendings, this will help align total spendings on agricultural research with the needs of the agricultural sector. An aggressive R&D investment strategy is absolutely vital for agricultural growth in an internationally competitive environment.

REFERENCES

Evenson, R.E. and D. Jha, "The Contribution of Agricultural Research System to Agricultural Production in India", Indian Journal of Agricultural Economics, 1973.

Evenson, R.E., "Technological Distance and Rice Productivity in India", *Paper for the National Workshop on Agricultural Research Evaluation: Methods and Applications*, 1993.

Kumar, P., C.C. Maji and R.K. Patel, "Returns on Investment in Research and Extension: A study on Indo-Swiss Cattle Improvement Project, Kerala", Indian Journal of Agricultural Economics, 1977.

5

Biotechnology in Horticulture Research

Agriculture is a way of life for more than sixty per cent of India's population. The cultivation of land not only sustains their livelihood but also provides a social milieu for their day-to-day living. No wonder the hopes, despairs, joys and sorrows of rural communities are woven around what the land provides. Around 35 years ago, agricultural production in India got a major boost with the introduction of dwarf varieties of wheat and rice. The introduction of these varieties led to a dramatic increase in the yields of the two crops. Some productivity enhancement came through the use of hybrids in corn, sorghum and millet although the area under cultivation of these crops has steadily decreased in the last decade. In the last 10 years the yields of rice and wheat have also plateaued out.

The productive agricultural areas in the North, due to continuous rice–wheat cultivation are encountering serious problems of sub-soil water depletion, deficiency of micronutrients in the soil and increase in the use of pesticides, fungicides and herbicides to control pests, pathogens and persistent weeds. Agricultural production is becoming more and more dependent on agrochemicals, thereby increasing input costs and causing significant damage to the environment and human health. Both farmers and consumers are at the receiving end –farmers by exposure to agrochemicals and consumers due to residues of agrochemicals in the consumed food.

While there is self-sufficiency in cereal grains at present, the yields and productivity of dryland crops, mostly grain legumes and oilseeds remain low and no major breakthroughs in productivity enhancement and yield stabilisation have been achieved. Currently, India is importing both grain legumes and edible oils to meet internal demands. India's population is expected to reach approximately 1.5 billion by 2050. It is estimated that around 300 million (roughly 30%) of India's population is suffering from malnutrition.

Due to socio-economic reasons, women and children are more at the receiving end in terms of micronutrient deficiency and low calorie intake. Nutritional security for everyone would require more extensive availability of grain legumes, edible oils, fruits and vegetables, milk and poultry products. The challenges of malnutrition, enhanced productivity and crop diversification can be met by better resource management and by breeding more productive, more nutritious and at the same time less resource input demanding crops.

Crop biotechnology, which broadly includes areas of development of transgenic crops, structural and functional genomics and marker-assisted breeding could provide us with the vital breakthroughs to achieve improvements in both quality and quantity in a sustainable manner. With the advent of techniques of genetic engineering in the early seventies, the natural barrier to gene exchange has been removed. Sequences from varied sources like bacteria, viruses and eukaryotic systems can be transferred to plants to develop transgenic crop varieties. Achievements, to date, in plant biotechnology have surpassed all previous expectations and with the development of high throughput instruments, the future is even more promising.

Recognising the importance and potential of biotechnology, the Government of India set up the National Biotechnology Board in 1982 which became a full-fledged department (called as Department of Biotechnology, DBT) in 1986 under the Ministry of Science and Technology.

A total of six such centres were set up initially at various Universities/Institutions namely, Jawaharlal Nehru sity (New Delhi), Madurai Kamaraj University (Madurai), Tamil Nadu Agricultural University (Coimbatore), Osmania University (Hyderabad), National Botanical Research Institute (Lucknow) and Bose Institute (Kolkata). A seventh centre was established at the University of Delhi South Campus in 1997. A total financial outlay of Rs 45.98 crores was made available for the development and research activities of CPMBs.

Most of the research on crop biotechnology currently underway in India is being supported by the Government and the DBT has been the main agency amongst government departments to support research endeavours in this area. The DBT has supported over the last 12 years a large number of research projects which deal with the development of *in vitro* regeneration and genetic transformation protocols of important crop species grown in India, the development of transgenics with genes of agronomic importance and several projects dealing with basic research in plant molecular biology.

To bring investigators with expertise in different facets of biology together to address defined research goals, a number of multi-institutional projects have been launched which include development of transgenics for resistance to geminiviruses in cotton, mungbean and tomato, resistance to rice tungro disease, resistance to bollworms in cotton, development of nutritionally improved potato with a balanced amino acid composition using a heterologous protein from *Amaranthus* and development of molecular methods for heterosis breeding.

Another set of projects include tagging of quality traits in rice and mustard, resistance to karnal bunt and rust in wheat and an extensive research network on rice biotechnology, which includes work on genomics, mapping, improvement in nutritional quality and resistance to biotic and abiotic stresses. Another multi-institutional project on tagging of quality traits in wheat by molecular markers has given some interesting leads. A number of markers linked to grain protein content, seed size, bread-making quality, etc. have been identified.

GENOMICS

Rice production in India holds the key to the food security of the country. After detailed consultations with experts at the National and International level, DBT and the Indian Council of Agricultural Research (ICAR) jointly took the initiative to participate in the international rice genome-sequencing programme (IRGSP) in June, 2000. Two laboratories in India, one at the University of Delhi South Campus and the other at the Indian Agricultural Research Institute (IARI) in New Delhi are jointly sequencing part of chromosome II. This region carries genes for a number of biotic and abiotic stresses and genes encoding some agronomically important traits. Around 15.58 Mb of DNA has already been sequenced till December 2002. A high quality draft sequence of rice genome by IRGSP has been declared completed by the Japanese Prime Minister Junichiro Koizumi on 18 December 2002 in Tokyo.

In order to take advantage of sequence information in rice, a workshop was organised in May 2002 to identify areas which should be funded for functional genomics of rice. It was decided to develop research progammes in the areas of:

(i) salinity and drought stress,

(ii) biotic stresses,

(iii) mapping of loci contributing to high yield.

The project on genome sequencing will also allow rapid progress in the characterisation of allelic diversity available in *Oryza sativa* and related species and genera for resistance to diseases and pests of rice.

In the 10th Five Year Plan the following areas will be given high priority:

— Rice genomics with particular emphasis on the study of genetic variability for achieving both yield increase and stabilisation in the crop.

— Molecular maps in rice, wheat and grain legume crops. To date, a molecular map with reasonable density (around 1000 AFLP markers) has been developed only for mustard (*Brassica juncea*). In wheat, tagging of some interesting traits has been accomplished. The work on cereal crops will continue and mapping work will be initiated in grain legume crops. Molecular maps will be used for mapping of QTLs for abiotic stresses, plant architecture, and yield and mapping of loci conferring resistance to pests and pathogens.

— Transgenics developed in various crops through DBT funding will be tested under appropriate containment conditions and if found useful will be subjected to nutritional and toxicological tests to determine their suitability for cultivation in the farmer's field.

— Research on development of pollination control methods for the production of hybrid seed will be further strengthened. Fixation of heterosis through apomixis will be another area of research.

— Particular emphasis will be given on an improvement in the nutritional status of some of our major crops. To begin with, the three target crops will be rice, wheat and potato. High-yielding rice varieties grown in India will be either transformed with constructs that would code for high β carotene content or by back-crossing these varieties to 'Golden rice' transgenics already developed in Switzerland. Improving the iron content in rice grains will be another thrust area. In potato, nutritionally enhanced transgenic Indian varieties of potato carrying the *ama1* gene are already under advanced field trials. If found suitable, these will be released for cultivation.

— Transgenic approaches to senescence retardation in fruits will be a major thrust area. It is important to reduce post-harvest losses in fruits and vegetables as under tropical and sub-tropical conditions senescence is much more rapid and energy costs of cold-storage are very high.

BIOSAFETY CONSIDERATIONS

Safety of transgenic material is a major concern. A major contribution of DBT has been to set up, in collaboration with other government departments, a framework for the evaluation and eventual clearance of transgenic crops for field cultivation. The release of transgenic crops is governed by the Indian Environment Protection Act (EPA) – 1986 which came into force from May 1986. The act provides a framework for the protection and improvement of environment. Later, the rules and regulations for the manufacture, use, import, export and storage of hazardous microorganisms, genetically engineered organisms or cells were notified under EPA 1986 on 5 December 1989.

A mechanism based on interaction between committees and different departments of Government of India has been set up. Such materials will have to meet with the approval of the following Committees: Institutional Biosafety Committee (IBSC), Review Committee on Genetic Manipulation (RCGM) and Genetic Engineering Approval Committee (GEAC).

Under the EPA 1986, the GEAC examines from the viewpoint of environmental safety and issues clearance or the release of genetically engineered organisms/ transgenic crops and products into the environment. The GEAC also grants permits to conduct experimental and large-scale field trials which are beyond the limit of 20 acres. In case of transgenic crops, applicants are required to seek clearance from the Ministry of Agriculture, which will accord the final clearance of transgenic crop varieties for agricultural activities.

Bt cotton was released for commercial cultivation in March 2002, following the biosafety procedures laid down by the Government of India. Three cotton hybrids have been granted permission for field sowing and are currently in the fields in six states namely, Maharashtra, Gujarat, Madhya Pradesh, Andhra Pradesh, Karnataka and Tamil Nadu. Transgenic mustard hybrids are also in advanced field trials and if found suitable, may be released for general cultivation.

ORGANIC AGRICULTURAL RESEARCH

Organic agriculture has become a widely accepted sector of agricultural and food research. In Europe publicly funded research has shifted towards organic farming, while research funding for conventional farming has been cut considerably. This paper takes a closer look at how organic farming research is organised in Europe and at what challenges researchers will face in the future.

Organic farming research has developed over four stages: first through pioneer farmers and scientists; then through pioneer private research institutes; through organic farming chairs at universities; and, finally, through organic farming projects at state research institutes.

In the very beginning, several scientists such as Sir Albert Howard, Lady Eve Balfour, Rudolf Steiner, Hans Mueller and Jean Boucher as well as practitioners like Maria Mueller, Raoul Lemaire and Maria Thun developed ideas for or carried out many research activities. Formal scientific research activities began in the 1970s. The following research institutions were founded within a very short space of time: the Research Institute of Organic Agriculture in Switzerland; the Louis Bolk Institute in the Netherlands; the Rodale Institute in the USA; the Ludwig Boltzmann Institute for Organic Agriculture and Applied Ecology in Vienna, Austria; the Elm Farm Research Centre in England; the Henry Doubleday Research Association (HDRA) with its research station in Coventry, England; and the Norwegian Research Institute for Organic Agriculture (NORSOK), now one of eight state institutes under the Norwegian Ministry of Agriculture.

Research Structures

Organic farming research is organised differently in the various European countries. Compilations of the most important research institutions are provided by Willer and Zerger and Padel.

Funding

According to recent reviews, the emphasis placed on organic farming research in a given country only partly reflects the density of organic farms. Leading in organic farming research are the Nordic countries: Denmark, Norway, Finland and, to a lesser extent, Sweden. Denmark has reserved 4-8% of its total research funds for organic farming projects. Organic farming research is also growing rapidly in Switzerland, where 3 to 4% of the total research budget of the Federal Office of Agriculture is allotted to the private Research Institute of Organic Agriculture, and another 2 to 3% are earmarked for organic farming projects at state research institutions.

In contrast, organic farming research is seriously under-funded in Italy, considering the density of organic farms there and the severe technical problems posed by the Mediterranean climate and horticultural crops. There is also a research deficit in other Mediterranean countries as well as in central and eastern European countries. In most other European countries, research funding corresponds to the proportion of organic farms to conventional ones.

Several research projects focusing on organic agriculture were funded under the EU's 2nd, 3rd and 4th Framework Programmes. The 5th Framework Programme explicitly calls for research on organic farming under Key Action 5 of the Quality of Life and Management of Living Resources Programme, which stresses the need for research on conversion to organic farming and for support for EU legislation. Five organic projects have been under negotiation since October 1999, which reflects a significant preference.

Private Research Institutes

Today, private research institutes continue to play a prominent role in Switzerland (FiBL), the Netherlands (Louis Bolk Instituut), England (Elm Farm Research Centre, EFRC and Henry Doubleday Research Association, HDRA) and Austria (Ludwig Boltzmann Institut). They act as main centres of competence in organic farming research or provide experts on organic agriculture for projects carried out at conventional research institutes, thus forming links between these and the organic sector .

University Chairs and Institutes

In some countries, organic agriculture research is mainly funded at the university level. Germany has several chairs for organic agriculture. Research on organic farming is also mainly conducted at the university level in: Austria (University of Vienna), Denmark (University of Copenhagen), the Netherlands (University of Wageningen), Sweden (Swedish Agricultural University) and Wales (Welsh Institute of Rural Studies, Aberystwyth).

This is also the case in the central eastern European countries of Hungary and Poland, where chairs for organic agriculture have been established.

The scientific requirements for Ph.D. theses and the flux of doctoral students make it extremely difficult to transfer academic results into agricultural practice.

Organic Farming Research Activities

Public research institutions specialising in organic farming exist in the following Scandinavian countries: Denmark (Danish Research Centre, DARCOF, Foulum); Finland (Research Station for Ecological Agriculture, Partala); Sweden (Centre for Sustainable Agriculture in Uppsala, CUL) and Norway (Norwegian Centre for Ecological Agriculture, NORSOK in Tingvoll Gard). With the exception of the Swedish CUL, which belongs to the Swedish Agricultural University, all centres are part of general agricultural research institutions. In Denmark, for example, the Danish Research Centre DARCOF in Foulum co-ordinates the research projects of 15 research institutes with 100 researchers. The Danish call DARCOF the "research centre without walls". At the moment, a follow-up programme (DARCOF II) is under evaluation, for which 22 million euro were allocated for 2000 to 2005.

In almost all European countries, minor organic farming research is taking place at universities and other research institutes not specialised in organic farming. In Italy and other southern European countries, organic farming research mainly depends on activities at such institutions. In Austria the state research institutes at Gumpenstein and Linz are conducting several research projects on organic farming. This is also the case in Switzerland, where organic farming research takes place at all six conventional federal research stations, chiefly at the Swiss Federal Research Station for Agroecology and Agriculture (FAL).

The French Institut National de Recherche Agronomique, INRA started to co-ordinate its organic farming research in 1999. The German national agricultural research station (Bundesanstalt fuer Landwirtschaftliche Forschung, FAL) will also intensify its activities in the future by setting up an organic agriculture research institute in the north of Germany (Trenthorst) in autumn 2000.

Since 1990, the UK Ministry of Agriculture and Fishery (MAFF) has financed 54 projects at universities, colleges and state or recently privatised institutes such as ADAS Terrington, ADAS Redesdale and IGER Trawsgoed. The Dutch government has also initiated research projects which are mainly carried out by conventional research institutes.

Co-ordination Activities

Because research in organic agriculture is now increasing and is therefore becoming more anonymous, it is very important that the efficiency of organic research and the exchange of preliminary results and project ideas be improved. In the Scandinavian countries, research activities are efficiently co-ordinated by the state-funded centres for organic agriculture, fostering a dialogue between the advisory services, practitioners and researchers.

In France the Institute Technique de l'Agriculture Biologique (ITAB) with its two regional organisations (the Groupe de Recherche en Agriculture Biologique (GRAB) in Avignon and the Groupement d'Agriculture Biologique du Nord, GABNORD in Lille) co-ordinates applied and on-farm activities. The Italian "Group of Researchers in Organic Farming" (Grupo di ricerca in agricoltura biologica, GRAB-IT) was founded in 1996. It aims to co-ordinate research efforts and organise workshops. In Austria the Forschungsinitiative Biologischer Landbau provides a forum for all institutions active in organic agriculture research. In Switzerland a working group comprised of members of FiBL and of the six federal research stations co-ordinates common projects. In England the Colloquium of Organic Researchers (COR) has met twice since autumn 1999. It aims to improve the dialogue between organic researchers.

Regional Networks

A scientific conference has taken place every two years since 1993 in the German-speaking countries. It is organised by an organic farming research institution in co-operation with the Stiftung Oekologie & Landbau (SOEL). In the Scandinavian countries, the Nordic Research Network for Ecological Agriculture co-ordinates teaching activities at the university level. The network discusses multidisciplinary research for the development of the organic farming system and publishes a newsletter.

The Mediterranean countries are beginning to co-ordinate their research activities within the IFOAM Mediterranean group Agrobiomediterraneo, which was founded in 1990 to meet the very special requirements of the Mediterranean climate.

Pan-European Co-ordination Activities

The EU has funded several concerted actions, such as the European Network for Scientific Research Coordination in Organic Farming (ENOF) and Documentation of Ecological Agriculture (DOCEA). The Network for Animal Health and Welfare in Organic Agriculture (NAHWOA) is a recent addition.

The FAO Regional Office for Europe (REU) has established a working group on "Research Methodologies in Organic Farming", a part of the SREN-ESCORENA network, which has met twice since 1998. ENOF has started to establish collaboration between all institutions working on the research, experimentation, demonstration or diffusion of organic farming techniques.

World-wide Co-ordination Activities (IFOAM)

The International Federation of Organic Agriculture Movements (IFOAM) promotes the dialogue between organic agricultural practice and research internationally. It has held thirteen scientific conferences since 1977.

Role of Internet

The internet has recently gained importance as a medium for the exchange of information on organic farming research. Most institutes have their own homepages where they present their activities.

FUTURE CHALLENGES

In the process of allocating more money to organic farming research, many national agricultural ministries and some international organisations have evaluated the needs and priorities of organic farming research. With far less than one per thousand of public and private research funding allocated to organic farming during the last 50 years, the results of such evaluations inevitably are long lists. Although organic farming benefits from basic and conventional research (e.g. resistance breeding, soil ecology, agri-environmental research, bio control etc.) most practical solutions suitable for integrated pest management or conventional farming can not be applied to the organic system.

Several factors have recently accelerated the need for research. The growth of organic farms has led to a diversification of production and of farm enterprises. Whereas in the past the main products were easy to produce under the restrictions of organic agriculture (milk, beef, wheat, selected vegetables), today products which pose challenges to the organic producers such as fruit, wine, difficult vegetables, oil-seeds, root crops and nonruminants like poultry and pigs have become progressively more important. The increasing marketing possibilities, now mainly via conventional supermarkets, are leading to a further acceleration of the demand for foods which are difficult to produce in close cycles and which depend on various inputs. Conventional retailers look for perfect appearance, high extrinsic quality, a constant daily supply and deliveries all year round. A third factor worth mentioning are the increasingly strict EU standards, especially in animal husbandry.

It can therefore be anticipated that future research will strongly concentrate on the direct regulation of phytosanitarian and veterinary problems and on affordable inputs into organic plant and animal production which comply with the organic standards (fertilisers, substrata, pesticides, fodder, feed additives etc.). A positive effect of this will be the replacement of problematic agents like copper. Scientists will also develop alternative therapies to antibiotics and anthelmintics for major animal diseases, both of which are very problematical from a human toxicological or ecotoxicological point of view.

On the other hand, systems-oriented research, which aims towards the prevention of acute problems, must be intensified; otherwise, the basic principles organic farming would be distorted. Therefore, it will become crucial to develop self-sufficient and self-regulating production systems for horticultural crops such as fruit, wine, vegetables, herbs and berries as well as for glasshouse production; and to better integrate non-ruminants into arable farms.

Consumer preferences have been shifting from fresh or dry products towards convenience foods. To differentiate between suitable and unsuitable processing and storage techniques with regard to organic standards, more research must focus on the quality issue.

Finally, organic farming researchers must protect the genetic resources necessary for plant and animal production and for food processing.

REFERENCES

Hoeoek, Karin, "Swedish Council for Forestry and Agricultural Research: Ecological Agriculture and Horticulture", *Research in Seven European Countries*, Stockholm, October 1997.

Krell, Rainer, FAO (ed.), "Biological Farming Research in Europe", *REU Technical Series No. 54*, FAO Regional Office for Europe, Rome, 1997.

Niggli, U. and Lockeretz, W., "Development of Research in Organic Agriculture", In: Oestergaard, (Ed.): *Fundamentals of Organic Agriculture*, IFOAM, Tholey-Theley, 1996.

6

Impact of Horticultural Research on Poverty

Assessing the impact of agricultural research on poverty is difficult, as there are so many ways in which agricultural research can have an effect. For this reason, many studies have tended to simplify the linkages between agricultural research and poverty and measure only one or two aspects of those linkages. While understandable, the approach can miss many important aspects of poor people's lives, including the diverse ways in which technology directly or indirectly affects their livelihoods.

SUSTAINABLE LIVELIHOODS APPROACHES

Sustainable livelihoods approaches have evolved from three decades of changing perspectives on poverty, how poor people construct their lives, and the importance of structural and institutional issues. The concept of "livelihoods" has become increasingly popular in development thinking as a way of conceptualising the economic activities poor people undertake in their totalities. The focus of development thinking in the 1970s on employment and "jobs" has given way to the realisation that while job creation in the formal sector continues to be one important strategy for poverty reduction, the reality for poor people in the South is that survival and prosperity depends on the pursuit of diverse and multiple activities simultaneously, by different family members, taking advantage of different opportunities and resources at different times. As Chambers wrote in 1997:

> They maintain a portfolio of activities. Different members of the family seek and find different sources of food, fuel, animal fodder, cash and support in different ways in different places at different times of the year. Their living is improvised and sustained through their livelihood capabilities, through tangible assets in the form of stores and resources, and through intangible assets in the form of claims and access.

Livelihood activities may be composed of, for example, year-round or seasonal formal-sector employment, informal trading or sale of labor, home gardens and food processing, livestock production, cultivation or use of natural or common property resources, labor exchange among family or neighbors, contracted "home work," borrowing, scavenging, stealing, and begging. They may be on or off farm, include local or international migration, involve elderly household members or children, be legal or illegal.

For poverty analysis and poverty reduction interventions to be effective, it is important to understand these multiple activities in order to understand the multiple sources of vulnerability faced by the poor, the multiple ways in which their lives are affected by structures and institutions, and the varied ways in which development interventions may strengthen or weaken these livelihood activities. In addition to recognising these activities, using livelihoods approaches requires an attempt to understand the processes that underlie poverty, and the social, cultural, political, and institutional contexts in which poor people live. Although the individual, household, and community are the primary levels of analysis, livelihoods approaches seek out the relevant interactions at micro, intermediate, and macro levels. Hebinck and Bourdillon point out the different ways in which a livelihoods framework is used in the field of development:

For policy makers... 'livelihood' provides a framework that focuses on poverty within the contexts of the people who are poor, and on the processes that underlie poverty. For consultants who operate in the field of development, 'livelihood' represents a framework for the formulation of development projects that focus on the people being affected by the project and the variety of ways in which they might be affected. For social scientists, such as anthropologists, sociologists and economists, 'livelihood' provides a framework for a holistic interpretation of the dynamics of development and the different rhythms of change. For plant breeders, soil scientists, and other technologists, the livelihood framework serves the purpose of linking their specific work and capacities with what people are capable of doing, what they are looking for, and how they perceive their needs. The livelihood framework thus provides a guide for research and intervention.

SUSTAINABLE LIVELIHOODS FRAMEWORK

The sustainable livelihoods conceptual framework is a particular form of livelihoods analysis used by a growing number of research and applied development organisations, including the Department for International Development (DfID) of the United Kingdom (one of its most ardent supporters), the United Nations Development Program (UNDP), as well as nongovernmental organisations (NGOs) such as CARE and Oxfam. It is primarily a conceptual framework for analysing causes of poverty, peoples' access to resources and their diverse livelihoods activities, and relationship between relevant factors at micro, intermediate, and macro levels. It is also a framework for assessing and prioritising interventions. The IFPRI/SPIA study is testing and adapting the sustainable livelihoods framework for use in agricultural research, with the aim of assisting agricultural researchers to conduct expost and ex-ante assessments of the impact of their interventions on poverty. To date, the vast majority of impact assessments in CGIAR centers have used conventional measures of poverty based on income and consumption data, and sometimes nutrition indicators. The sustainable livelihoods framework takes as a starting point an expanded definition of poverty that looks *beyond* the following:

— conventional poverty measures based on income, consumption, or nutrition to *additional aspects of poverty and well-being*, e.g., access land, water, credit, or education, vulnerability to natural disasters, political rights, phys ical safety, and social relationships that provide economic security and social well-being;

— "today's poor" to who is vulnerable or likely to be *"tomorrow's poor"*;

— aggregated household or head counts to *the effects of social differentiation* by class, ethnic group, gender, and other locally-specific social differences; and · external standards to *self-perceptions by local communities* on who is poor and what poverty means, taking into account what people themselves value.

One feature of this framework is that it looks at more aspects of people's lives than how many live on a purchasing power of $1.00 a day or how many households consume less than 2,000 calories per person per day. For example, participatory poverty assessments or case study research can identify the features by which people in rural areas themselves identify poor or well-off households.

A second key feature of the sustainable livelihoods framework is that it recognises people themselves, whether poor or not, as actors with assets and capabilities who act in pursuit of their own livelihood goals. While this may seem obvious, in many cases the poor have been regarded as passive victims or recipients of government policies and external aid.

The conceptual framework for sustainable livelihoods is intended to be dynamic, recognising changes due to both external fluctuations and the results of people's own actions. The starting point is the vulnerability context within which people operate. Attention is next given to the assets that people can draw upon for their livelihoods. Assets interact with policies, institutions, and processes to shape the choice of livelihood strategies. These, in turn, shape the livelihood outcomes, which are often the types of impact we are interested in. However, those outcomes are not necessarily the end point, as they feed back into the future asset base.

The vulnerability context encompasses

— *trends* in population, resources, and economic indicators such as prices, governance, or even technology;
— *shocks* such as changes in human or animal health, natural disasters, sudden economic changes, or conflict; and
— *seasonality* in prices, agricultural production, employment opportunities, resource availability, or health.

Vulnerability here refers to things that are outside people's control. It is usually negative but it can also provide positive opportunities. It is not objective "risk" that matters, but people's subjective assessments of things that make them vulnerable. These matter because both perceived and actual vulnerability can influence people's decisions and hence their livelihood strategies. This is especially important for whether people are willing or interested in adopting agricultural technologies. The asset base upon which people build their livelihoods includes a wider range of assets than are usually considered. Rather than looking only at land or other classic wealth indicators, the sustainable livelihoods framework suggests consideration of an asset portfolio of five different types of assets:

— *Natural capital* includes land, water, forests, marine resources, air quality, erosion protection, and biodiversity.
— *Physical capital* includes transportation, roads, buildings, shelter, water supply and sanitation, energy, technology, or communications.
— *Financial capital* includes savings (cash as well as liquid assets), credit (formal and informal), as well as inflows (state transfers and remittances).
— *Human capital* includes education, skills, knowledge, health, nutrition, and labor power.
— *Social capital* includes any networks that increase trust, ability to work together, access to opportunities, reciprocity; informal safety nets; and membership in organisations.

Though most versions of the sustainable livelihoods framework are limited to these five kinds of capital, some add political capital as a sixth type of asset, which can include, for example, citizenship, enfranchisement, and membership in political parties—all assets that can be key in obtaining or operationalising rights over other assets.

Policies, institutions, and processes affect how people use their assets in pursuit of different livelihood strategies. This box on the diagram refers to both formal and informal institutions and organisations that shape livelihoods by influencing access to assets, livelihood strategies, vulnerability, and terms of exchange. They may occur at multiple levels, from the household to community, national, and even global levels. The public and private sectors, civil society, and community institutions may all be relevant considerations; laws as well as culture can also be included.

All of these influence people's livelihood strategies, i.e., the choices they employ in pursuit of income, security, well-being, and other productive and reproductive goals. As discussed above, what is important about the livelihood strategies approach is that it recognises that households and individuals may pursue multiple strategies, sequentially or simultaneously. This means that, even in the context of agricultural research, we should not assume that someone is automatically a "farmer," or that people with other businesses are not involved in farming. Nor should we overlook even small livelihood strategies, because they can be very important, especially for the poor, who often pursue many livelihood strategies either to make up enough income or to provide a measure of security. The pursuit of multiple activities can have important implications for cash and labor availability at different times of the year, and for the relevance of specific development interventions for poverty reduction.

Livelihood outcomes encompass many of the types of impact of interest for the study of the impact of agricultural research on poverty. Potential outcomes include conventional indicators such as income, food security, and sustainable use of natural resources. Outcomes can also include a strengthened asset base, reduced vulnerability, and improvements in other aspects of well-being such as health, self-esteem, sense of control, and even maintenance of cultural assets, and thus have a feedback effect on the vulnerability status and asset base.

The sustainable livelihoods framework draws on a number of theoretical and conceptual approaches to development thinking; in this sense it is more a holistic and synthetic framework than an entirely new set of concepts. What the framework does is provide a method for thinking about the multiple and interactive influences on livelihoods without overlooking important explanatory factors. In this respect, it provides a "checklist" of issues to be considered in designing research initiatives or program evaluations. Not everything on the checklist can be included in one study, so prioritisation is necessary. The framework provides the advantage of allowing researchers to understand the parameters of the "big picture," and then narrow the scope of the study to what can have the highest impact or what is most relevant to the important stakeholders (including researchers). The framework may guide researchers to consider and prioritise less visible factors and local priorities that may or may not revolve around production and consumption or even physical or financial resources, but could instead relate to education, safety, or legal rights.

The framework may also identify issues that are highly salient in explaining livelihood impacts, but are either (1) too far outside of the domain of the research to be a focus of study, e.g., vulnerability to domestic violence in the context of a study of agricultural research; or (2) while relevant to the study, are impossible to include, e.g., a high level of political violence that is either too sensitive or dangerous to address. In these cases, the framework can help make explicit what is *not* included, but may still be important to understanding chains of causality or important constraints on the ability of a technological intervention to affect livelihoods. Use of the sustainable livelihoods framework involves acknowledging complexities that can be hard to manage in a study. Rather than shy away from this complexity, use of the framework implies a willingness to acknowledge that livelihoods—and the process of affecting them—are complex. It also implies making an effort to achieve the most comprehensive understanding of these issues possible.

The framework requires researchers to think holistically, not just about certain types of assets such as land and credit, but also about the potential interaction of five or six kinds of assets, and the complementarities between assets and their sequencing. For example, membership in a social group (social capital) may be necessary for access to rights (political capital) and land (natural capital), which is necessary for access to credit (financial capital), which, in turn, is needed to purchase inputs to take advantage of a new technology. This understanding may lead to a different choice of intervention. The framework can also provide a structure for thinking about conflicts between livelihood objectives, e.g., whether increased production might conflict with human capital development or protection of the natural resource base, or whether income maximisation through increased cash crop production might increase vulnerability of women through decreased production of crops used for small enterprises.

Livelihood Framework and Agricultural Research

The sustainable livelihoods framework brings in many considerations that are often not included in an impact study dealing with agricultural research. At the same time, it may not be obvious how agricultural research and technologies might fit into this framework.

Agricultural research and technologies can reduce vulnerability, such as when irrigation reduces susceptibility to fluctuations in water supply, or pest control technologies reduce vulnerability to crop or animal loss. However, research and technologies can also increase vulnerability, such as when new varieties are more susceptible to crop failure if conditions are not right, or farmers have to purchase the seed again every year under conditions of cash constraints. In Bangladesh, the susceptibility of polyculture fish production to disease was cited by rich and poor alike as a serious problem. In Mexico farmers find that improved varieties of maize have advantages in terms of height and yield but are more susceptible to rotting and certain pests. Drought in the early 1990s was a particularly severe source of vulnerability in Zimbabwe, affecting farmers who had adopted hybrid maize through widespread crop loss, fertiliser burn, and loan defaults. Richer farmers who had used increased incomes from higher maize yields to diversify into cattle were better protected from these drought shocks. The importance of *perceived* vulnerability is illustrated in the Zimbabwe case in terms of the extent to which people perceived themselves as vulnerable to magic and witchcraft—seen as a significant factor explaining good and bad yields.

Research and technologies are strongly linked to the asset base. The bestresearched aspects of this relationship are the types of assets that are required to adopt new technologies. For example, much of the debate on the Green Revolution centered around whether or not large landholdings (natural capital) were required to adopt the various components of the green revolution package. Considerable policy emphasis has also been given to expanding agricultural credit (financial capital) and roads or transportation (physical capital) in order to permit technology adoption. Human capital, in the form of knowledge and skills, is often required to properly make use of many new technologies. It is now increasingly recognised that social capital can facilitate adoption of technologies that operate on a large spatial scale, wherein collective action is needed to coordinate the action of individuals for common investment or adherence to rules.

Many natural resource management practices, including integrated pest management, community nurseries, rangeland management, irrigation, forestry, and watershed management fall into the category of technologies that are facilitated by collective action. In the case study of fish ponds in Bangladesh, one NGO disseminated the technology to households that had private fishponds, which meant that natural capital (rights to land with a fish pond) were required for adoption; this favored relatively well-off households. In another study district, an NGO assisted in the formation of groups of landless or land-poor households that could collectively rent a pond, and then disseminated polyculture fish technology to the group s.Here social capital, as embodied in the groups, became a way of overcoming natural capital constraints. However, success of the technology then hinged upon maintaining cooperation, and many households lost out when disputes erupted within the group.

Further illustrations from the SPIA case studies of the relationship between agricultural technologies and access to assets can be found in Mexico, where poor farmers say they do not plant the improved maize varieties because they perceive them as more management intensive, requiring more fertiliser and labor (financial and human capital constraints). Similarly, in the case study in Kenya, land and labor availability were cited as constraints on adoption of ICRAF soil fertility replenishment (SFR) technologies. Although adoption of SFR technologies were seen as increasing yields and improving soil quality, leaving land fallow, as required for some of these technologies, is difficult for resource-poor farmers. This can potentially be a source of increased inequality if better-off farmers are able to take advantage of the technology and achieve higher yields than can poorer farmers. Access to assets is, in turn, affected by the vulnerability context, where vulnerability to shocks such as AIDS and other illnesses reduces labor availability.

Even technologies that are designed to be adapted to less favored areas and poor farmers may not do so without farmers' access to certain assets. The Zimbabwe case study has found that although the HYV maize was designed to do well in drought prone areas, farmers in better agroecological zones adopted faster than those in middle and poor zones. Preliminary results indicate that this is because in these zones, farmers had more agricultural capital stock and livestock to protect them from risk, and had better marketing channels. Assets and people's access to them also have many gender dimensions, affecting the value of technologies to men and women.

In Zimbabwe, men have been found to prefer the improved varieties, while women seek out the openpollinated varieties. This is because women have less access to the credit and cash required for certified seed and fertiliser. Moreover, women use their social networks to acquire open-pollinated seeds and find these do well with cow manure. Women also have less access to formal maize markets where improved maize is sold, an example of how policies and institutions influence technology adoption choices and consequences among different social categories of farmers.

Agricultural research can shape the asset base as well. This is most easily seen in new equipment that becomes part of physical capital, or irrigation or soil fertility management practices that improve the natural capital of water and land. As noted above, for example, in Zimbabwe, increased yields and incomes from HYV maize enabled some farmers to diversify into livestock. Participatory or action research processes can strengthen the human and social capital asset base when knowledge is generated and groups are formed to work together on the research. The knowledge generated through NGO training programs in fish or vegetable production in Bangladesh was a clearly valued addition to local human capital.

Social capital plays an important role in influencing impacts of agricultural technology, because of the ways in which social networks and social relationships facilitate and constrain technology dissemination. Pre-existing social capital may be used in dissemination, or new forms of social capital may be developed for the purpose of dissemination. The latter is the case in Western Kenya, where ICRAF and its partners, as well as several NGOs and the Ministry of Agriculture and Rural Development, have developed new local groups and innovative methods for dissemination of technology, information, and training among and between villages.

In Mexico, many poor farmers prefer to obtain seed through their informal social networks of family and friends, and prefer to learn from their neighbors' successes and problems, rather than through official government channels. In Zimbabwe, social capital plays a role in the relationship that rich and poor male farmers growing HYV maize have developed with marketing institutions, as well as networks they use for information and technology dissemination. Women farmers obtain open-pollinated varieties and market maize through their own social networks. For Zimbabwean households, higher yields from HYV maize strengthen reciprocity relationships with kin and friends, especially urban-rural relationships: the ability to send maize to the city ensures that other resources are sent back to the country, suggesting that the well-being impacts of social capital can depend on the ability of both parties to bring resources to the table.

Social capital, perhaps more clearly seen as complex social relationships, can also interact with new technologies and produce negative consequences, or otherwise complicate dissemination efforts. For example, in the ICRAF Western Kenya study, social status in the village influenced dissemination processes in a number of ways. The perceived low status of the local farmer chosen as an extension agent sometimes contradicted peoples' expectation that innovation comes from external and superior sources. Furthermore, which farmers attend training and dissemination meetings depends on who in the village makes the invitations, and the political alignment and social standing of other invitees. While SFR technologies give new status to farmers who are seen as "successful farmers" and attract visitors to the village, strong social tensions have been created where some farmers are seen as "favored" by external organisations, exacerbated by "clanism" and politics.

Finally, the policies, institutions, and processes (PIP) dimension of the SL framework is relevant to the study of agricultural research in several ways. Agricultural research and technologies can be considered as a component of the PIP that composes the environment in which farmers operate. They are institutions whose interventions change people's options in pursuing their livelihood strategies. By changing the relative returns to different factors or assets, technologies can effectively change the distribution of the value of assets within and between households.

For example, aquaculture technologies allow people to make flooded land produce fish for income and home consumption, while research on drought-tolerant crop varieties can allow farmers without irrigation to pursue farming with less fear of disastrous losses. In Bangladesh, modern rice varieties have increased labor productivity and led to better labor contracts for the landless, with not only higher real wages but also more dignity. Productivity increases also affect income, which may have other complex impacts on institutional arrangements, such as by creating demand for other products, which, in turn, creates more employment opportunities through forward and backward linkages.

Agricultural research also interacts with other political institutions at national, regional, and local levels, and is facilitated or constrained by public policies. Land tenure arrangements, legal rights to natural capital, marketing institutions, input packages, prices, and other policies influence the ability of farmers to take advantage of technologies, and influence rich and poor farmers, and men and women farmers differently. In Zimbabwe from the early 1980s, farmers were only allowed to plant HYV maize; they were provided with input packages and credit, and marketing services. These had mixed consequences: it had positive impacts on yields, but failure of these varieties and damage from fertilisers during drought led to loan defaults, and women suffered decreased control over maize crops and income.

Vulnerability factors may lead people to change the seeds used, such as in Mexico where farmers have, by design and accident, promoted crosses between improved and local maize varieties. Some of these "creolised" varieties are perceived to be more resistant than the improved maize to drought and pests—and require less labor at critical times—hence reducing these farmers' vulnerability. The vulnerability context and asset portfolios of farmers should lead agricultural researchers, including farmers themselves, to adapt technology to these contexts. Moreover, the organised efforts of farmers (social capital) as well as individual farmer's experimentation, adaptation, and innovation can lead to changes in technology that better meet the needs of resource-poor farmers. Finally, policies, institutions, and processes can shape the decisions of agricultural researchers in technology development, e.g., through commodity markets and prices, laws and policies related to land and water, education and extension, as well as direct financial investments in technology at the national level.

The sustainable livelihoods framework is not the most direct way of examining the impact of agricultural research, because it requires consideration of so many intervening factors at multiple levels, factors that may or may not have a major impact and may or may not relate to the technology. But this reflects reality. After all, agriculture is only one part of people's livelihoods, and agricultural research and technologies may affect only one part of the total farming system, as illustrated in Figure 1. Understanding the other factors that impinge at each point can be critical to improving the ultimate impact of agricultural research.

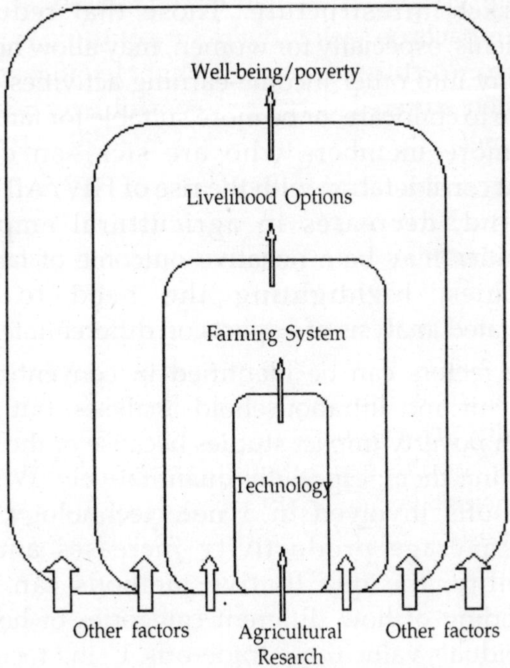

Figure 1: Role of agricultural research and technology in livelihoods and well-being

Experience with the IFPRI/SPIA project has shown that, although use of the sustainable livelihoods framework can appear daunting, using it to assess the impact of agricultural research is both manageable and helpful in suggesting relationships to examine. Employing this expanded conceptual framework can change the findings on the impact of agricultural research. Technologies that stabilise yields in the face of climatic or other fluctuations, thereby reducing vulnerability, may emerge as more valuable for improving people's livelihoods and well-being than technologies that maximise average production, but with higher fluctuations. Technologies that do not require many purchased inputs may be more accessible to households with low income or access to transportation and market infrastructure. Those that reduce labor requirements, especially for women, may allow households to diversify into other income-earning activities or devote more time to childcare, or be more suitable for families with one or more members who are sick—an especially important consideration with the rise of HIV/AIDS. On the other hand, decreases in agricultural employment opportunities may be a negative outcome of laborsaving technologies, highlighting the need to conduct disaggregated analysis of impacts on differentiated groups.

These factors can be identified in conventional farm management and intrahousehold analyses, but are often omitted in poverty impact studies because of the difficulty in measuring them, especially quantitatively. Where there are trade-offs involved in a new technology, such as between average productivity increases and higher vulnerability, the qua litative methods can improve understanding of how different categories of households and individuals value those trade-offs. Using focus groups and other qualitative methods is also useful in identifying factors that might otherwise be overlooked, or to prioritise which of the many potential effects are important for poor people in that area. In the Bangladesh case studies, labor power (the most basic form of human capital) emerged as the critical dividing line between the poor and the extremely poor.

Therefore, technologies that improve the health of family members can be very important, while those that increase the risk of disease could deepen poverty. Similarly, vegetable cultivation that could be undertaken by women around their homestead emerged as having a more important role for poor families than conventional economic analysis would indicate, not only because of the contribution to household nutrition. Additional benefits identified through the sustainable livelihoods analysis include:

— strengthening social capital linkages through sharing of "high-status" vegetables with friends and relatives;
— production in the homestead site, so that women could be economically active without leaving the home and being vulnerable to harassment or violence; and
— empowerment of women as they learned to deal with traders who came to buy vegetables.

Using a common conceptual framework facilitates comparison of results across case studies. The IFPRI/SPIA case studies were originally developed with a variety of different research questions, conceptual frameworks, and research designs. When the sustainable livelihoods framework was adopted across the five cases, the first step was to look at how the original questions "mapped" into the sustainable livelihoods framework. Many issues that had been identified related to either aspects of vulnerability, assets, intervening institutions (e.g., extension services), livelihood strategies (e.g., employment opportunities created), or outcomes. In this way, the studies will demonstrate how to contextualise a common set of concerns and arrive at an assessment of the types of outcomes to which the conceptual framework points. The crosscutting themes that have been identified include

— *Dissemination pathways.* How do different types of research and dissemination pathways affect poor people's ability to influence research priorities and access research outcomes? How do poor people view, respond to, and come to trust different pathways?
— *Social differentiation.* How does social differentiation mediate the impact of research on poor people? This includes intrahousehold (gender, age, relationship,

etc.) and interhousehold (class, religion, caste, ethnicity, agricultural or pastoral, etc.) dynamics.

— *Assets.* How does poor people's capacity to access and manage assets affect their ability to access research outcomes/technologies and convert them into positive livelihood streams?

— *Institutions and processes.* How do different policies, institutions, and organisations mediate poor people's ability to access research outcomes/technologies and convert them into positive livelihood streams?

Comparability across case studies is also addressed by a set of guiding principles across case studies, including

— adoption of a shared understanding of core concepts, including poverty, assets, livelihoods, and vulnerability;

— application of a common conceptual framework for assessing the likely impact of agricultural research on poverty; and

— commitment to applying interdisciplinary perspectives, combining qualitative and quantitative research methods, and integrating the economic and social analyses into a unified study.

Limitations of the Sustainable Framework

Although it is too early to take stock of lessons learned from using the framework and its analytical potential, certain limitations have emerged in applying it at the design phase. One objective of the SPIA project is to test the use of the framework in the context of agricultural research and adapt it to increase its usefulness. In designing the SPIA case studies, we have identified aspects of people's lives that are not captured in the framework, and yet are important to explaining people's decisions and choices, and consequent livelihood outcomes.

One aspect is the notion of "culture." How things have been done in the past, the relationship of certain crops or practices to ancestors, or their importance in festivals can influence whether people adopt a new crop or related farming practice, or whether they value the traits of that new crop. In this sense, the framework could be expanded to include "cultural assets," which would include, for example, beliefs, traditions, language, identity, festivals, and sacred sites.

Other aspects that can be attributed to culture may include taste and texture of agricultural products, as well as status that is associated with certain varieties of crops or values that determine how certain resources should be used, or the age or gender appropriateness of cultivating certain kinds of crops or involvement in the management or marketing activities that accompany them. These cultural assets or factors may not have direct economic value but are centrally important in people's lives, choices, and well-being. They can have economic value as well, such as where assets are transformed into tourism or handicraft production. In one village included in the Mexico SPIA study, sewing was an important activity for the purpose of keeping the village supplied with dresses for its almost monthly festivals.

There is something unsettling, however, about trying to fit these aspects of culture, identity, and values into an "assets" or "capital" framework, and it may not be worth doing so. Hebinck and Bourdillon take exception to the livelihoods framework's overreliance on the notion of "capital," because of the overemphasis that this places on the material aspects of people's lives:

> One of the problems is that it is an economic metaphor that does not do justice to the nature of people's activities, which are not entirely oriented towards material gain. Although material gains are a very important aim in the notion of livelihood, 'livelihood' does not span only the commoditised world and associated values The term also incorporates the

> noncommoditised, nonmaterial, and cultural part of life and sets of values that are embedded in local cultural repertoires...for example... community values determine how and under which conditions forest resources should be used, and how the rights to these resources are embedded in culturally defined relationships.

They also draw on Long's work on knowledge to critique its inclusion conceptualisation as part of "human capital," because as such it implies that knowledge is a universal, culture- neutral resource that can be accessed as a commodity, rather than recognised as a social construct or a relationship that is redefined within a local context.

Also missing from the sustainable livelihoods framework are the notions of power and power relationships. There is a large body of literature on these topics, for example, on intrahousehold power relationships and women's empowerment, and the extent and nature of women's power as well as its increase or decrease can have a powerful influence on livelihoods. A new agricultural technology may increase or decrease women's decisionmaking power; participation in a group involved with villageto- village technology dissemination may increase members' confidence, which involves material and nonmaterial aspects of improved well-being. Class-based power imbalances may lead to technology access, or the extension needed to benefit from the technology, being captured by richer farmers. Lack of political power may mean that poor farmers do not have access to certain marketing channels, or cannot get sufficient prices for their crops. As Ashley and Carney note, "the SL framework overall can convey a somewhat cleansed, neutral approach to power issues.

This contrasts starkly with the fundamental role that power imbalances play in causing poverty." Institutional aspects of power can be captured through the PIP and social capital dimensions of the framework, and empowerment can be identified among livelihood outcomes. However, this needs to be made explicit where these factors are considered important to a given study. Also, whereas the household can be seen to be one of the "institutions" to be considered within PIP, and intrahousehold power relationships a part of this, individuals and class-based social groups are not institutions and it is difficult to see where individual, class-based, or ethnic group-based power dynamics or conflict suggest themselves within the SL framework. There is also a question as to how "politics" fits within the framework. In one district in Zimbabwe, farmers have been wary of new technologies because they view the phase out of older varieties and their replacement with new varieties as a conspiracy between Agritex officers and the privatesector distributor to discredit the government. This argues for the interpretation of PIP to include political party politics as part of institutional processes, but more broadly for the incorporation of an explicit notion of politics to capture these significant dynamics, even where they are not related to formal institutions.

Finally, the framework does not lead one to incorporate the importance of historical factors. For example, where people have had a history of problems with external interventions, this can influence their reception to new interventions. In the Zimbabwe study, we are exploring the ongoing influence of farmers' bad experiences with loans taken in conjunction with the adoption of earlier varieties of improve maize. In focus groups conducted as part of the Mexico SPIA study, some villagers reported that they distrust government agents and programs, and thus the advice they give on modern maize varieties, preferring to learn from their neighbors' successes and problems. They also said they trusted more the seeds they acquire from friends, family, and neighbors.

Poor people in Bangladesh also reported that government agencies served the rich rather than the poor, and tended to prefer NGOs, except in one village where a bad experience with an NGO had also eroded trust in such organisations. This illustrates the continuing significance of historical social relationships in contemporary development initiatives. In all five case studies, we are considering people's "trust" of different dissemination pathways. Trust normally encompasses a historical dimension, where trust in a dissemination institution is based on past experience, whether related or unrelated to the institution in question.

More generally, the framework diagram suggests a snapshot approach, when, in fact, it should be seen as dynamic. Livelihood strategies, vulnerability factors, asset portfolios, and policies, institutions, and processes are often in a state of flux, so it is important to operationalise the framework in a way that incorporates this temporal dimension.

The above critiques are addressed by including these additional concepts wherever they are relevant. Thus, use of the sustainable livelihoods framework does not have to be limiting; it is simply not sufficient on its own for pointing to all possible factors relevant to a study or development intervention. It must be used in conjunction with concepts, tools, and modes of analysis that have long been used in other fields, such as development sociology, anthropology, and history.

Sustainable livelihoods analysis needs to draw on a set of "tools" that may include gender analysis, institutional appraisal, stakeholder analysis, market analysis, and others. The framework does not explicitly address the differential conditions, assets, and strategies of socially differentiated groups. Therefore, additional explicit attentio n must be given to the implications of gender, ethnicity, class, or other types of social differentiation. It is likely that in a given community, livelihoods analyses will need to be conducted for different social groups, and sometimes even at the level of the individual rather than the household. Even key concepts often emphasized as "sustainable livelihoods principles," such as the idea that "poor people themselves should be key actors in identifying and addressing livelihood priorities", are not explicit in the framework. Instead, participatory poverty assessments are another tool that is recommended for use when using the sustainable livelihoods framework. In fact, as Ashley and Carney point out, there is "no explicit mention of poverty in the SL framework" and that the assumption that use of SL approaches will contribute to poverty elimination must be continually tested.

Research Methods

The prospect of covering all aspects of the sustainable livelihoods framework, or even all questions identified as critical for assessing the livelihood impact of agricultural research, can be daunting. Especially for econometric analysis, the way in which so many factors are interrelated creates endogeneity problems that would require ever larger data sets to resolve. What is required, then, is an integrated, interdisciplinary approach that draws upon both quantitative and qualitative data collection and analysis.

By piecing together data from a variety of sources, it is possible to cover key aspects of the sustainable livelihoods framework relatively efficiently. Rather than seeing this as a second-best solution, such a combined approach can actually provide a more convincing analysis than any single method. This is because studies have found that people respond differently to quantitative and qualitative information. Numbers are required to convince some audiences, while others will be unimpressed by numbers, but relate more to in-depth and contextual information gathered using qualitative techniques. The fact that livelihood activities are so varied, and often intermittent, "small" or noncommoditised means that surveys are likely to pick up some activities and miss others. Providing examples from their work in Zambia, Norton, Owen, and Milimo argue that:

> Most aspects of rural livelihoods are not captured in either income or expenditure-based survey data. This is because they are neither commoditised nor evident enough to the researchers to be allocated 'imputed values'...Energy (fuelwood) and herbal medicines are two examples. A significant element of the 'safety net' for rural people in times of stress consists of 'famine foods' that can be gathered from bush and fallow lands.

Triangulation and crosschecks on the results of different methods can improve confidence in the overall study. Use of quantitative and qualitative methods provides a richer base for analysis, where data from each method help to interpret the other. In the IFPRI/SPIA study, the major data-collection methods used include surveys, focus groups, key informant interviews, in-depth household case studies, and secondary data. The case studies combine social and economic (as well as some biophysical), qualitative and quantitative, participatory, and conventional (or extractive) data. Although there is often a tendency to equate social, qualitative, and participatory data collection on the one hand, and economic, qua ntitative, and extractive data on the other, the studies also collect quantitative social information, qualitative economic information, and use both participatory and extractive methods for each.

All the SPIA poverty impact case studies include household surveys. Some have panel data for the same households over a number of years, which allow for analysis of changes over time. Some of the surveys (notably on vegetables and fish in Bangladesh, and maize in Zimbabwe) have collected data at the level of the individual household member, which allows for comparison between men and women, and also helps to capture the full range of livelihood strategies within the household. Sampling to cover the range of wealth/poverty categories is critical for the surveys. Although some qualitative data are included in the surveys, researchers analyze most survey data using econometric techniques.

All the case studies make use of focus groups to elicit collective experience and opinions. Separate groups are convened for men and women of different wealth/poverty categories. For example, in Bangladesh, six focus groups are held in each selected village (men and women separately for the very poor, poor, and nonpoor categories of households). Preexisting survey data helps in the disaggregation of wealth groupings for the focus groups, particularly in communities where a wealth ranking exercise may be divisive or difficult to carry out (e.g., because of large community size or time limitations that prevent researchers from getting sufficiently acquainted with a community to comfortably carry out such an exercise). Where possible, households that are selected for the surveys are included in the focus groups to improve the comparability of the information obtained by the different sources.

During the focus group meetings, a range of participatory and extractive data collection activities is conducted: seasonality mapping, identification and ranking of livelihood activities and sources of vulnerability, as well as discussions of the technologies being studied and dissemination approaches. In some of the studies (e.g., Kenya, Zimbabwe, and Mexico) focus groups are used following a series of household case studies to further investigate issues raised, check whether the findings resonate or contradict, and receive feedback on the research findings. In other studies (especially in Bangladesh) focus groups are the primary means of qualitative data collection, but are followed up with in-depth interviews or case studies of individuals who participated in those groups. Analysis of focus group data is done partly by the respondents themselves, partly by the field staff's summaries, and partly by the lead researchers on each case study team.

Key informant interviews allow the research team to follow up in more detail with individuals that have specialised knowledge. This may include researchers from CGIAR and national centers, NGO, community organisation, or government project staff, extension agents, local seed distributors and shops, agricultural researchers from the private sector, community elders, chiefs, early adopters, etc. Semi-structured interviews allow the researchers to go in with a core set of information that they hope to collect, but also to follow up on relevant topics that emerge during the course of the discussion. This information is especially important to address the policies, institutions, and processes affecting the research. As with the focus groups, analysis is shared between the respondents and lead researchers.

In-depth household case studies provide more detail on the complexity of household livelihood strategies, particularly in the Kenya, Zimbabwe, and Mexico cases. Researchers live in sample villages for three to six months, spending time in the homes of a subsample of the survey households, conducting informal interviews, observing and participating in their daily activities, such as farming, extension field days, and social interactions and activities. Such participant observation can provide insights that are not available from other methods and inform and refine the questions asked in other, more structured, data collection. Qualitative data are coded according to issues identified by the research questions in advance, as well as issues that emerge through the fieldwork. These data are analysed by the research team, taking into account findings of the focus groups and surveys.

Secondary data sources from government and other researchers' studies sho uld not be overlooked. These can provide the basis for sampling frames, cross-check the information from the study with other regions or nationally representative samples, and even provide direct information for the existing study. For example, participatory poverty assessments (PPA) conducted in Bangladesh provided the basic criteria for classifying households according to poverty or wealth status. These criteria could then be applied to the households in the survey, and used to select participants for the focus groups. In Mexico, where earlier ethnographic studies of communities included in our case study region exist they provide important historical information on livelihoods, institutions, and processes. In Kenya, previous and ongoing studies carried out in the same region as the case study provide additional information on related technologies, asset portfolios, gender issues, cultural issues in adoption, social networks, and other issues.

Explicit attention must also be given to linking the different sources of data. Depending on the sequence of data collection, insights from the surveys might be followed up in the focus group or key informant interviews and participant observation, or vice versa. In most of the cases an iterative approach to data collection is used. However, in each of the case studies, the quantitative and qualitative data take a somewhat different relationship to each other and provide different types of interpretative power. In the Kenya and Mexico cases, pre-SPIA qualitative studies informed small surveys. The SPIA projects then initiated new qualitative work (household case studies and focus groups) in a wider set of communities (and included new issues), which were informed by the earlier work. Larger scale household surveys have been designed drawing on the findings of the new qualitative work, and household level case studies will provide a depth of understanding that will be used to interpret findings of the surveys.

In Mexico, qualitative work explored reasons for people's preferences for different maize varieties, and the main risk factors that they face. The importance of these risk (or vulnerability) factors and perceived advantages of maize characteristics identified in this qualitative work will be tested quantitatively through a survey of a wider sample of farmers. Case studies will deepen our understanding of how different maize characteristics respond to this vulnerability context, as well as issues such as people's perceptions and trust of the pathways through which seeds enter communities (whether by government channels or informal social networks) and how this influences people's choices.

The Bangladesh case on vegetable and fish technologies used an iterative process of survey data collection and qualitative data collection on intrahousehold dynamics and women's empowerment followed by another round that collected individual- level indicators of empowerment. The focus groups built upon the analysis of the survey data, e.g., looking at why, if the technologies were profitable, households had not expanded their vegetable or fish production and why household incomes of adopting households were not necessarily higher than for non-adopters. Key informant interviews can the n follow up on problems with the program that focus group members identified. Careful attention was given to sampling for the surveys, focus groups and household case studies, with links between the samples. For example, efforts are made to include the same households in the surveys, focus groups, or in the surveys and household case studies. Thus, the quantitative and qualitative data can lend each other the maximum interpretive power.

One of the most challenging aspects of such multi-method research involves assembling a research team with the proper mix of skills. Each of the IFPRI/SPIA studies has a case study leader (usually an economist) who is an international staff member of the lead CGIAR center, with extensive experience in the case study country. Although technical scientists are not formally part of the research team, they often serve in informal advisory capacities and in some cases assist with aspects of the study. The case study leader is paired with an international social analysis team member (a sociologist or anthropologist) with extensive experience in the region and is involved in at least one other case study, which increases comparability across cases. Each case also has national economics and social analysis experts who guide the data collection and analysis, and who work with teams of less experienced researchers, engaging in training and capacity building. The field staff require strong analytic and facilitation skills in order to conduct the focus group and household case studies, while the key informant interviews are often conducted by the national or international social or economics experts. Finally, an external Advisory Committee composed of a leading economist, social analysis expert, and livelihoods expert advise the project on best practice methods, integration of the research, and dissemination strategies.

Working with interdisciplinary groups from the international to the local level has been a valuable learning process in mixed method research and in integrating economics, sociology, and anthropology, and can provide a model for strengthening the capacity for future CGIAR research to address poverty. The challenges of developing such mixed teams lie in ensuring that members can communicate across disciplinary lines, respect each other's contributions, and find the time to integrate the findings or insights from the other members into their own work. The sustainable livelihoods approach helped in this integration because it provided a common framework into which each member could contribute but also required the contributions of others. It also helped that all members of the team were familiar with the situation of poor people in rural areas, and the livelihoods framework drew on concepts that they were familiar with from their experience in the field.

The significant contribution from DfID for social analysis sent a strong signal of the increasing recognition among the international development community of the importance of interdisciplinary and mixed method research. The additional budget that the social analysis has for data collection as well as analysis also means that qualitative data collection is not seen as taking resources away from survey data collection, but is a valued addition to the project. On the other hand, large ly separate budgets for the economic and social analyses can also lead to separation rather than integration of the components, requiring more vigilance for integration. The coordination of activities, especially between people based at different institutions, remains a challenge, which could be reduced if all members of the team were within one institution or could focus only on this project. Yet drawing from different institutions can also be an important aspect of capacity-building. Ultimately, however, interdisciplinary work within each institution would be strengthened by having more disciplines represented within the research staffs.

REFERENCES

Ashley, C., and D. Carney, *Sustainable livelihoods: Lessons from early experience*, London: Department for International Development (DfID), 1999.

Chambers, R., A. Pacey, and L-A Thrupp, *Farmer first: Farmer innovation and agricultural research*, London: Intermediate Technology Publications, 1989.

Kerr, J., and S. Kola valli, *Impact of agricultural research on poverty alleviation: Conceptual framework with illustrations from the literature, Environment and Production Technology Division Discussion Paper 56*, Washington, D.C.: International Food Policy Research Institute, 1999.

7

Economic Impacts of Horticultural Research Programmes

The goal of most public agricultural research organisations is to undertake research and development work that will ultimately improve the productivity and sustainability of the agricultural and food sector. In today's world of scarce public funding and greater accountability, governments, donors and research managers are increasingly demanding assessment of the economic returns to their investments in research.

Agricultural research organisations (AROs) world wide undergo various forms of reviews and evaluations as is appropriate to their role as public institutions. Formal economic impact evaluations, which form an increasingly important element of these evaluations, are carried out for both internal and external clients of the ARO. The first group includes governments and donors who increasingly require accountability for the funds they invest. Since agricultural research is one of many competing investment alternatives in the portfolio of national governments and international aid agencies, these funding agencies need stronger and clearer evidence of the net social benefits of their investments in research.

The second type of client includes scientists and research managers within the ARO who require information on economic impacts to provide feedback to their research programs. By generating a better understanding of how technology influences the welfare of agricultural producers and consumers, impact evaluation can improve targeting of research programs and help adjust resource allocation across programs.

Although the same studies can serve both types of clients, there is inevitably a tension between the two. In the current climate of declining funding, more emphasis has been placed on the first objective to demonstrate to external clients the positive payoffs to research investments. However, these studies tend to focus on 'winners' and are less useful to research managers in reallocating resources or adjusting programs. Few research organisations evaluate unproductive programs, although such studies may be very valuable from the point of view of enhancing research efficiency. Institutionalisation of regular evaluation of all applied research should be the long-run goal of AROs.

Economic impact evaluations can be classified into ex ante evaluations and ex post evaluations. Ex ante evaluations are undertaken before the project or program is initiated as an aid in priority setting and ex post evaluations are undertaken after diffusion of a research product has been initiated, to assess actual impacts on the ground. However, ex post evaluations generate information that is useful for the selection, planning and management of future research programs, such as plausible adoption paths.

Ex post impact studies range in scope and depth of evaluation from simple story-telling and anecdotes, to partial and comprehensive assessment of economic impacts. Evaluations of the "story-telling" types are based on researchers' description and assessment of the importance and value of results from their research. Often these descriptions are supplemented by backof-the-envelope calculations of economic benefits. Such evaluations may provide decision makers with pertinent information but their lack of rigor often undermines their credibility, especially in today's climate of accountability.

Partial impact assessment studies quantify the application of research results but do not estimate aggregate benefits. Adoption studies to trace the use of innovations from the research station or on-farm trials through networks of adopters, are the most popular type of partial impact study. Adoption studies may also evaluate *private* benefits in the form of increased farm production and incomes, assess client satisfaction with research results, and seek to understand why a technology is or is not being used. As such they are especially useful in feeding back information to research decision makers—that is, research managers, stakeholder advisory committees, and scientists.

More comprehensive impact studies look beyond adoption information to quantify the effectiveness of research in terms of achieving the major objectives of AROs, especially the enhancement of productivity. This is quantified in impact studies by estimating the economic benefits to both producers and consumers produced by research, computing a rate of return (ROR) to the research investments. There are two broad approaches to estimating a ROR—the econometric approach that relates productivity changes to investment in research, and the economic surplus method that builds benefits from the bottom up, based on estimated productivity changes at the field level and adoption rates for each technology.

Comprehensive studies may also trace wider economic benefits of research through factor and product markets. Also AROs are increasingly being asked to address other objectives beyond efficiency, such as equity and poverty alleviation, environmental quality, food safety, and nutrition. Undertaking a comprehensive economic impact assessment requires considerable financial resources and specialised skills in economics. Developing-country NARSs and increasingly the IARCs, are especially short of economists and other social scientists to carry out impact assessment studies. Research managers, therefore, need to weigh the tradeoffs between costs of more comprehensive analyses, and the value of the additional information that they provide.

SYSTEM-WIDE PROGRAMME OR PROJECT

Impact assessments can be carried-out at different levels of aggregation—individual research projects, specific research programs, or the research and technology system as a whole—depending on the objectives of the exercise. There are a number of arguments against conducting impact assessment of individual projects. First, it is often not possible to assess a project's impact independently of other projects within the research program. Second, the idea of doing impact assessment is to assess whether *total* investment in research has paid off in terms of national goals. In any large portfolio of research projects, there are bound to be a wide range of returns, including many failures. The incidence of some "dry holes" should not deter investment since a few profitable strikes more than cover the costs of the unsuccessful ones. Third, the costs of evaluating each project is usually high in relation to the size of the project investments, and selecting single projects for impact evaluation usually leads to selection bias towards known success stories. However, large, discrete projects in some research systems, such as some mega-projects in the IARCs, may warrant individual studies of impact.

The most appropriate level for impact evaluation is at the program level, which includes costs of all successful and unsuccessful projects and may involve evaluation of one or more products of the research program. These evaluations usually use the benefit-cost analysis framework to measure the rate of return or net present value of research program investments. Program-level evaluation should be a major focus of impact assessment studies and should be conducted on a regular basis (e.g., once every five years).

Evaluation at the system level is less frequently performed. This requires evaluating the agricultural R&D system at an aggregate level (country or state) or by specific sub-sector e.g., crops versus livestock). System-level evaluation studies are resource- and dataintensive, requiring time-series data and the use of econometric techniques. Results are most useful for those who fund research and less useful for decision making on resource allocation within the research organisation.

SPATIAL AND TEMPORAL DIMENSIONS OF THE ANALYSIS

Impact assessment studies must necessarily define the time period and spatial dimensions of evaluation. The time period of the analysis depends on the main objective of impact assessment. It is sometimes interesting to analyse impacts over a long period of 20 or more years into the past. However, if impact analysis is to be used in decision making by research funders or implementers, it should focus on technologies released in the most recent 5-10 year period. This inevitably involves projection of the benefit stream into the future, since adoption and diffusion will likely be incomplete. Thus, useful ex post impact studies will normally include some element of ex ante estimation of adoption.

The spatial dimension of the analysis also needs to be carefully determined. This involves answering questions such as what is the geographic mandate of the research program being assessed? National research is usually funded based on political boundaries (a state or country), although impacts will follow agro-ecological boundaries and socio-economic characteristics. Most analyses assess direct impacts within the politically defined constituency of the funding body, which will often be less than total impacts due to spillovers across political boundaries. In the case of internationally funded research, political boundaries are less important, although spillovers to the industrialised world may not be counted as impacts by donors who emphasize poverty alleviation and food security.

ECONOMETRIC APPROACHES TO IMPACT ASSESSMENT

The econometric approach employs a production function, cost function, or a total factor productivity analysis to estimate the change in productivity due to investment in research Econometric studies of research impacts are conducted within the framework of a production function that incorporates conventional inputs (land, labor etc.), non-conventional inputs (education, infrastructure etc.), and the stock of technical knowledge (e.g., investment in research and extension). Research effort is used as one of the argument in the function and is usually measured by lagged research expenditures.

Recent efforts have expanded the specification to include resource quality variables (e.g., soil erosion, nutrient status etc.), and weather variables. The estimated research coefficient(s) (measuring marginal product) is then used to calculate the value of additional output attributable to the lagged research expenditures (holding other inputs constant) and to derive a marginal ROR to the research investment.

There are many technical areas of debate and refinement in the literature on econometric methods, such as the use of a production function, cost function, or total factor productivity (TFP) index, functional form, the length and shape of the time-lag, the appropriate method of determining the rate of return from the estimations, and the quality of indices used as the dependent variable. However, the main constraint on the wider application of econometric approaches in developing countries is data availability and quality.

The econometric approach requires good-quality time-series data, which are difficult to obtain below the national or state level in most developing countries. Therefore, the approach is generally best for ex post evaluations of entire agricultural research systems over a long time period (at least 25-30 years), if the quantity and quality of data allow the use of statistical methods. Much of the work in this area in developing countries has been pioneered by Evenson.

The approach is less relevant for individual research organisations, either national or international, since time series data are rarely sufficiently long enough or complete enough or available at the needed level of disaggregation to allow econometric estimation. Also because of the long time period of analysis, results are dated and therefore less useful to decision makers.

One compromise is to use panel data to capture both cross-sectional and time series variability. Secondary data of a panel nature are increasingly available for many of the variables at the district level, especially production and input data and some recent studies have even included district-wise data on resource quality. Especially for analysis of impacts of natural resources management (NRM), the appropriate level of analysis is at the cropping or farming systems level defined in terms of a relatively homogeneous agro-ecological resource base. For example, Murgai, Ali and Byerlee find dramatically different patterns of TFP growth across cropping systems, underscoring the importance of aggregation criteria.

Increasingly time-series panel data are becoming available at the household that could be employed in econometric approaches to impact evaluation. For example, Cassman and Pingali use panel farm-level data to estimate TFP for intensive irrigated rice systems in the Philippines. As panel data become more widely available, the use of econometric approaches to research evaluation will expand, although the emphasis will continue to be on providing historical aggregate returns to investment in research to those who fund research, rather than to inform decisions of research managers in AROs.

ECONOMIC SURPLUS APPROACH

Economic surplus has been the most popular and fruitful approach used to assess the consequences of investments in agricultural research since the pioneering study by Griliches. The approach estimates returns on investment (generally, a weighted average rate of return over time) by (1) calculating the change in consumer and producer surpluses that results from technological change brought about through research that causes the industry supply function to shift outward, and (2) using estimated economic surplus together with research costs to compute the net present value (NPV) or internal rate of return (IRR). Economic surplus analysis requires information on values of production affected by research, yield increases, reduction of unit costs and/or the maintenance effects of research, adoption rates, research and adoption lags, cost of R&D, and the appropriate discount rate for converting benefits and costs into present values or for comparing them against an IRR. Price elasticities of demand and supply, and the magnitude and nature of the supply shift determine the relative benefits gained by producers and consumers. In practice there is wide variation in the process, and the methods and data used to measure research costs and benefits. The following sections summarise good practice at each step in the process.

Measuring Programme Costs

Although measuring program costs may appear straightforward, most past studies have underestimated costs for a number of reasons. First, program costs should include all research, development, and technology transfer costs. Research costs include not only direct program costs but also research overheads, such as administration expenditures, depreciation of capital assets, and costs of complementary services such as libraries.

There is also the difficulty in knowing how to treat the input of past knowledge, especially basic and strategic research, given that research lags are a critical determinant of the returns on the investment. Usually, impact analysis excludes such sunk costs associated with previous-generation technology development.

One issue that frequently arises at the national level is whether to include development costs incurred by an IARC that had a role upstream in generating the technology outputs. In the past, IARC products have generally been regarded as a free good by NARSs in impact assessment.

However, with NARSs paying an increasing share of the budget of IARCs, a trend that is likely to accelerate, good practice is to include pro-rated IARC costs according to the country's share in the production of a commodity. For evaluation of IARC research, NARSs' cost should almost always be included. Finally, technology transfer costs, including the cost of on-farm verification, public expenditures on extension, and expenditures by the private sector (e.g., farmer organisations), should be included.

In defining the costs, it is often difficult to appropriately delineate program-specific costs. unless an accounting department of a research institute is very sophisticated. In practice, approximations have to be employed, such as allocation of scientists' time by program and using an institute-wide average cost per scientist. For certain categories of costs there are other definitional and measurement problems that make the cost estimation step less than straightforward. For example, markets for scientific skills are neither efficient nor free of distortions and shadow pricing may be appropriate.

Evaluation of Research Products

Impact assessment requires the definition of the products of the research organisation. Most commonly the product is new or improved knowledge that may be embodied in a production input, such as seed. New knowledge may be provided directly to the user in the form of information (e.g., a crop management recommendation). Most impact studies have emphasized embodied knowledge, but since a considerable share of the output of research organisations is in the form of information, this type of product is treated below.

Estimating the K-factor

The central piece of information for any research evaluation study using the economic surplus method is the shift in the supply curve that has resulted from research (K-factor). Most commonly K is conceptualised as a vertical (downward) shift in the supply curve. A more conservative assumption is to use a pivotal shift in the supply curve.

A number of options are available for estimating K, depending on the purpose of the analysis, data available, and the overall methodological approach being applied in the study. Some studies have estimated commodity-supply functions directly, with past expenditures on research included as an explanatory variable in an econometric model. Some studies estimate production functions and deduce the value of K from the estimated production function shifter.

These methods, however, require comprehensive time-series data on inputs and outputs. Hence most ex post economic surplus studies deduce the pattern of past supply shifts (K-factor) attributable to research based on $K_t = \alpha_t k_t$: where (a) α_t is the adoption rate of the technology in year t, and (b) k_t is the per-unit cost reduction in year t resulting from technological change.

Estimating the adoption of program outputs

Both the temporal and spatial dimensions of the adoption parameter are important in determining the estimates of actual shifts in the supply function (the K-factor) over time. Of these the most difficult is to obtain estimates of temporal parameters, such as when technology adoption begins, when it reaches its peak, and when disadoption begins. The shape of the lag distribution and the lag length in the earlier years is relatively important in determining economic benefits. The S-shaped logistic curve has been widely used following Griliches and is generally an acceptable approximation to reality. In empirical work, simpler linear functions, polynomial lags, or trapezoidal lags have also often been used and some allow the modeling of disadoption as well—a limitation of the logistic model.

Adoption data are provided by farm surveys, agricultural statistics, seed and other input sales, and estimates from research and extension workers. For some technologies, especially those related to livestock research, the tools of GIS are also used to extrapolate the distribution or occurrence of adoption- inducing factors (such as prevalence of pests) in a region. In an ex post setting, the data and information on the year in which a technology was released and a few estimates of adoption at selected years in the past can be used to estimate the adoption curve and the appropriate time lag. For discrete technologies, such as crop varieties, retrospective questioning of farmers on the year of adoption provides acceptable estimates of the adoption curve. However, for adoption of research products that are released as information on continuous variables (e.g., rates, timing, and levels of inputs), adoption information is best collected over time.

Per unit reduction in costs (k-value)

The size of the research-induced change in the unit cost of

production—the k-value—is a crucial determinant of the total benefits from research. The k-value is usually estimated by the yield increments per unit area or animal resulting from a technology. Translating a yield increase into a cost saving must, however, be adjusted to reflect changes in input use and take account of the industry supply elasticity. At the firm level, a research-induced per-unit cost reduction may be realised as a result of: (a) the cost reduction arising from changes in productivity when input use is held constant at the preadoption optimum, and (b) the cost reduction associated with changes in the input mix to optimise input combinations under the new technology. The measure of research-induced cost savings therefore needs to reflect research-induced changes in the use of inputs and their opportunity costs. At the level of the individual firm, this means that estimated per unit cost savings ought to reflect the comparison of the commodity enterprise budgets between the new and old technologies to take account of both effects.

The results of on-farm trials, especially farmer-managed trials, and other experimental data are essential for estimating the k value for a specific technology since they control for all other variables. Results from one-off experiments over several locations provide estimates of k for specific agro-ecological zones, while trial data over several years can be used in a statistical model to deduce yield-growth rates attributable to research. In livestock research, simulation models are often used to derive estimates of productivity impacts of a technology.

A critical step is to translate increases in experimental yields into industry-level cost savings. Experimental yields are typically higher than farm yields and gains in experimental yields often exceed gains experienced on farms. But the sizes of these differences vary by location and technology, and it is difficult to make empirical generalisations. In practice, a conservative approach is to scale down experimental yield gains to better reflect likely onfarm gains.

Historical data on farm yields over time and in different locations are often more readily available and more complete than experiment data. However, due to the multiplicity of factors affecting yield changes, these data are not generally a good source for estimating kvalues. However, with a carefully specified statistical model, the historical record can be very informative and useful when combined with, and juxtaposed against, experimental data.

Deriving the aggregate supply shift (K-factor)

Even with good estimates of the per unit cost reduction (k-value) and the adoption rate of the technology, there are additional methodological problems in deriving the aggregated value of the supply shift due to a research program. One difficulty relates to the possibility of *aggregation bias* if different program components yield very different time profiles for cost savings (i.e., k-values). In is usually best to disaggregate the measure of k by individual technologies, to account for different time profiles, so long as they can be regarded as independent.

For technologies with a high degree of complementarity or substitution, the only alternative is to treat them jointly and estimate the k-value using budgets for the alternative packages of technology.. Another conceptual challenge facing impact assessment at the program level is to determine the *counterfactual*—what would have been the shift in the supply curve in the absence of research. Without research, some changes in productivity may occur due to spill-ins from other research programs, or alternatively, productivity may decline due to depreciation in the knowledge stock.

Large errors, usually overestimates, are possible from a lack of rigor in thinking about the "with" and especially, the "without" situation. One approach is to model research only as an acceleration of adoption, with benefits derived only from shortening the time lag. The major challenge is to estimate the without-research adoption curve. The effects of maintenance research on the estimated incremental research benefits also need to be carefully assessed. Thus, in order to ensure that the estimated incremental net benefit represents the benefits of the research investment under consideration, one needs to carefully construct the possible "without" scenarios.

A final issue is whether to treat the products of research as *discrete or continuous* outputs. Generally for economic surplus analysis, technologies are regarded as discrete products and the impact assessment of a program considers all technologies released in a given time period. However, in some situations, and especially for varietal technologies in post-Green Revolution areas, research outputs are continuously released, adopted and replaced. In these cases, k is better conceived as an annual average rate of change in unit cost, based on say the rate of genetic gains in yields achieved through release of new varieties, rather than a one time shift.

In sum, good practice for calculating the K-factor includes;

1. Combine technical, scientific, and economic information from a number of sources. If available, use the results of previous experimental trials to assess changes in cost and yield.
2. Reflect research-induced changes in the use of inputs and their opportunity costs in the measure of cost savings.
3. Disaggregate supply-shift estimates by components of a research program.
4. Carefully construct the "with" and "without" scenarios to ensure that the estimated incremental net benefit

truly represents the benefits of the research investment under consideration.

5. Most importantly, make explicit the assumptions applied, so that others can transparently assess the analysis.

Tradable Status, Output Prices, and Partitioning of Benefits

Total economic benefits and the partitioning of benefits among producers and consumers requires knowledge of the supply and demand curves for the product. The key issues are to determine the tradable status of a commodity, a gap in much of the research evaluation literature, and the appropriate price for valuing outputs.

Many analysts have uncritically assumed the extremes of a completely open or closed economy. Liberalisation of markets and trade in recent years means that many commodities that were non-tradables are now tradable, implying that more, and perhaps most, domestic research benefits are now being captured by producers. However, in situations of poor infrastructure or long distances between producing and consuming points, there is a large "wedge" between import- and export-parity prices so that potentially-tradable commodities, even basic grains, are converted into non-tradables within a wide price range.

Figure 1 summarises a common market situation for food staples in developing countries. Urban consumers in a port city with demand curve Du face a supply curve, S^oS^u dominated by imports at price, P_I, while rural interior producers and consumers effectively operate in a closed economy in the price range, P_I +t to P_I, where t is the transport cost from rural areas to the port. S^oS^r and D^u are the supply and demand schedules for the rural interior market, respectively. In this situation, it is possible, and even likely, that with a research-induced shift in the supply curve from S^o to S^1, rural consumers register a gain in consumer surplus as they move from R^o to R^1 while urban consumers at U^o experience no such change.

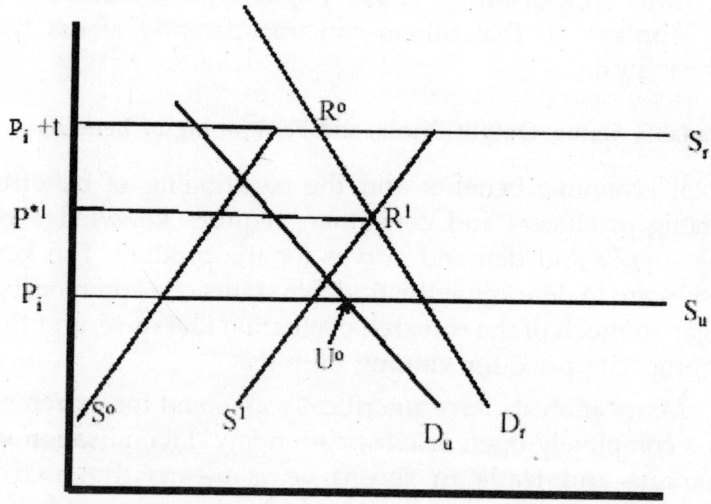

Figure 1: Effect of Interior Transport Costs on Tradeable Status in Evaluation of Research Benefits for Consumers

Thus where most of the food-purchasing population are located in urban areas, and consume tradable food staples, it is likely that consumer price effects of research will be limited. This characterises the situation of many countries in Latin America, North Africa, and increasingly much of Asia. On the other hand, where most of the population is located in rural areas, and infrastructure and markets are not well developed, markets for food staples in rural areas are likely to operate in the non-tradable zone in a wide range between P_i and $P_i + t$ in Figure 1. This is the case for much of sub-Saharan Africa. The main point is that the tradable status of a commodity needs to carefully assessed for each situation. Even within a country, a food staple will often be non-tradable in one area (e.g., rural areas with poor infrastructure) and tradable in another area (e.g., urban areas located on the coast).

Another important market condition to consider in estimating economic benefits is possible presence of policy distortions in the form of taxes or subsidies on output, exports or imports, production quotas, and production target prices, that affect output prices. Depending on the size of the output market and the trade status of the country, the presence of market distortions can significantly complicate the analytical model for estimating research benefits. Alston, Norton and Pardey is a good source for different formulas that can be used for computing changes in economic surplus under different market conditions and policy interventions. The extent of any distortion, if present, needs to be measured to derive an "economic price" to be used in the estimation of economic benefits. Byerlee and Morris provide conditions under which border prices or estimated equilibrium prices are appropriate economic prices, in the presence of policy distortions.

Total benefits are relatively insensitive to assumptions about the elasticity of demand and supply. Indeed, if the purpose of the economic analysis is to measure only the total economic surplus and not its distribution between consumers and producers, then a simplified benefitcost approach gives reasonable estimates of economic benefits. The simplified approach is valid in cases where the country is clearly a "price-taker", and where the intervention being evaluated is not expected to change the country's status from being a net importer to net exporter of that commodity. The advantage of this approach is that it is simple and presents a more transparent view to non-economists of the data and assumptions used in the analysis.

Key elements of good practice are:

1. Carefully study the market situation of the commodities affected by research, and determine an appropriate tradable status for estimating total surplus.

2. Use economic prices that are appropriately adjusted to reflect policy distortions in the output market. All monetary variables should be expressed in real terms.
3. Estimates of research benefits should be disaggregated by commodities, production environment or geographical basis if the parameter estimates are different for different components of a research program.

Calculating the Return to the Investment

The final step for the analyst is to bring together all the information to calculate the NPV or IRR of research investment. A number of spreadsheet templates have been developed specifically for economic surplus computation—MODEXC originally developed by Lynam and Jones, RE developed by the ACIAR, and DREAM developed at ISNAR/IFPRI.

There are several options for evaluating the stream of net benefits—NPV, IRR and Benefit-Cost (BC) ratio. The NPV method is preferred where concern over the scale of investment is not an issue. More commonly, the IRR is computed and compared to the opportunity cost of funds, although it does not reveal either the scale of the investment or the value of the programs. The best approach is to use a combination of IRR and NPV calculations to summarise the relevant information on the total returns to research.

This basic economic surplus approach can be extended in many ways. Substitution effects on the production and consumption side can be readily incorporated if a full set of supply and demand elasticities, including cross elasticities is available. To incorporate agro-ecological differences and regional impacts, geographical information systems are increasingly being applied to spatially characterise and map research or problem 'domains' within inter-regional models.

Good practice in data analysis and calculating summary measures include:

1. Creating a spreadsheet template for each of the commodity/products/markets affected by a research program using product- (or zone-) specific parameter values for each disaggregated research component (and respective spreadsheet template).
2. Documenting the logical structure of the model/method used for calculating the benefit and cost streams and clearly laying out the assumptions and sources used to derive model parameters.
3. Calculating both summary measures (NPV and IRR) to indicate the profitability of research investment.
4. Carrying out appropriate sensitivity analyses by varying some of the critical assumptions made in the model and to check the robustness of the results.

BROADENING THE APPLICATION OF IMPACT ASSESSMENT

Most economic impact studies have addressed commodity-specific research and, within that category, there has been an overwhelming emphasis on varietal technologies. For other types of research, especially research on natural resource management (NRM), productquality, crop and livestock management research, and social sciences, specific conceptual and methodological difficulties arise in applying the economic surplus framework. Additionally, some research by AROs is aimed at modifying institutions, and some generate externalities (positive and negative) which complicate the measurement of economic surplus. This gap is now being filled, especially with impact assessment studies being conducted by CGIAR centers such as ICLARM, ILRI, ICRAF, CIFOR, and IFPRI.

Impacts of Management Research

Many of the products of research are not embodied in

tangible inputs such as seed, but are provided as information, in the form of a recommendation. Adoption of information research products is often difficult to track, complicates attribution to research, and raises thorny problems of valuation.

A good example is crop and resource management research, where little economic impact work has been carried out, although this type of research accounts for about one-half of all crop research in developing countries. Information from crop management experiments is usually summarised in the form of production recommendations, which define rate, timing and methods for using inputs, as well as the conditions under which these recommendations apply. The value of improved input management information depends on the interaction of input response with location-specific climatic and field conditions (e.g., integrated pest management or phosphorus maintenance doses conditional on soil test information).

It is especially difficult to measure benefits of improved information from research compared to what would have occurred in the absence of the program. The effects of research must be separated from other sources of information, including farmers' learning-by-doing and private sector suppliers. Often this is approximated by the yield and cost differences observed between adopters and non-adopters of a management practice—that is, the area between the "with" and "without" adoption curves. Traxler and Byerlee provide a five-step framework for assessing benefits in such situations:

1. Identify research areas for which an improved management practice has been supplied as a new recommendation (i.e., new information) issued to farmers.
2. Determine which practices farmers have modified in a manner consistent with the new recommendations.

3. Determine whether a revised recommendation has caused the change in farmers' practice.
4. Measure the impact of each research-induced change in crop management practice on economic surplus, and
5. Sum the economic surplus across practices and compare the benefits stream to the costs of research and extension.

A refinement of the above approach is to track changes in farmers' subjective beliefs about payoffs to a practice in response to improved information provided by research. This is empirically more challenging but has been successfully applied by Pingali and Carlson in a commercial agricultural setting.

Social Science Research

Assessing the impact of stand alone social science research has a number of parallels with crop management research—i.e., social science research that is intended to be used directly by farmers and policy makers rather than as an input into other research programs. This includes social science management research that leads to more rapid adoption of technical innovations, marketing research that improves price incentives to farmers, and policy research that leads to an increase in agricultural productivity due to the removal of policy distortions.

Determination of the relative contribution of the social science research information to the changed policy or management practice is clearly central to impact evaluation. In recent years, IFPRI has initiated and commissioned several studies to develop methods for estimating the impacts of social science research in general and food-policy research in particular. Methods proposed or used for evaluating policy research range from a cross-country econometric approach, calculating the national income impacts of timesavings resulting from hastened policy decisions, valuing information and calculating the savings in dead weight loss of policy distortions, to simple case studies documenting and qualitatively assessing the results, outcomes and impacts of research. However, attempts to evaluate research projects in this area indicate that more experience is still required to define good practice in this area.

Impacts on Natural Resources and the Environment

Attention to effects of research on the natural resource base is relevant for both productivityenhancing research, such as commodity research, to the extent that some of the gains are made at the expense of reductions in the quality of the resource base, as well as to research directed explicitly at enhancing natural resources and environmental quality, or at least reducing their degradation. NRM research outputs are not only often highly location specific but monitoring the multiple bio-physical consequences of NRM research outputs in the field is expensive, and the interplay between the economic and bio-physical domain is strong.

Moreover, there are still many conceptual and methodological issues that need to be resolved in impact assessment of NRM research. Assessing the full range of impacts, requires that appropriate information sources be tapped at various levels, such as field, watershed, and regions. It also raises technical difficulties of measuring and valuing resource degradation, (e.g., soil loss under alternative management practices), or environmental pollution (e.g., through inappropriate use of agricultural chemicals).

Measuring and valuing changes in resource quality

There appears to be a consensus that estimates of the *quantities* of resource depletion or environmental damage are easier to get (although often costly) than estimates of the *correct prices* to attach to the quantities. Even in valuing *private* shadow prices for onsite effects of resource quality, there are valuation issues such as whether to use foregone output, replacement costs, user costs, or option and existence value to measure the costs of resource degradation.

In the simplest cases, such as soil nutrient 'mining', costs can be approximated by its replacement cost (i.e., the market price of chemical or organic fertilisers needed to fully replenish the nutrient). Likewise attempts have been made to value soil erosion effects by estimating erosion losses and output foregone using simulation models of soil run off and crop productivity.

Problems are compounded with externalities where costs of agricultural chemical pollution may include regulatory costs, health-related costs, and unpriced environmental costs. A number of techniques now exist for placing economic values on environmental changes, including health effects of pesticide poisoning, effects on soil degradation and contingent valuation of water pollution. A final complication with valuing changes in natural resources and environmental quality is the effects of policy distortions, such as subsidised inputs that foster overuse of chemicals, or institutional policies such as land tenure that affect adoption of improved management practices.

Conceptually, such assessments can and should be included in a broader economic impact analysis of agricultural research. The main problem is that they are data-intensive and require highly developed skills in collecting and statistically manipulating large quantities of data. For these reasons, it is unlikely that valuation of natural resources impacts will become an integral part of research impact studies in developing countries in the near future, although as we argue later, IARCs should be providing leadership in developing cost-effective and robust methods for wider inclusion of these impacts.

Resource quality in TFP

There is a parallel literature on the inclusion of resource quality variables and environmental externalities in econometric approaches using TFP. Joint outputs or externalities of agricultural production such as soil erosion, depletion of soil fertility, and groundwater aquifer mining are usually not taken into account in TFP measurement yet are clearly central to evaluating research impacts on natural resources. To address these problems, several researchers have argued that changes in the quality of the natural resource base should be accounted for in a total social factor productivity indices (TFP). A few studies have attempted to empirically measure Total Social Factor Productivity (TSFP), usually using data from experimental plots.

The inclusion of the costs of environmental regulation and abatement, as well as benefits of improved environmental quality, in TSFP is the subject of considerable debate. There are both conceptual and practical problems in using TSFP. On the conceptual side, TFP indices are based on an underlying production function, assuming profit maximising behavior. Therefore, inclusion of resource degradation is valid only if degradation is on-site and under the possibly dubious assumption that farmers are aware of, and take into account its long-run effects in their input and output choices. The most relevant case for agriculture is the accounting for soil nutrient mining and the on-site cost of erosion, since these are probably internalised by farmers in decision making. If it is a pure externality (off-site cost) that does not affect the system being evaluated (e.g., water quality for consumers), it should not enter the calculations, although private regulatory costs may be included as an input into production.

TSFP also requires that inputs and outputs be valued at long-term *economic* prices. Again the profit maximising assumptions underlying TFP require that observed input and output choices have to be valued at the prices faced by producers. Therefore, valuing production at the societal shadow prices of goods rather than producer prices is conceptually flawed. In addition, as discussed above, valuing resource quality and particularly externalities is especially challenging.

In sum, the recent emphasis in the economics literature of searching for an all embracing single measure of sustainability in the form of TSFP has not been helpful. Rather attention should now turn to doing a better job of measuring productivity and trends in resource quality and relating the two through production function analysis. Inclusion of resource quality variables in the production function is also an advantage in that it allows valuation of resource degradation through the estimated marginal product.

Non-research Products of Research Organisations

Another ARO product that is difficult to fit into an "economic surplus" model of impact assessment is the economic evaluation of the performance of an ARO in non-research activities such as training, networking, and advisory services aimed at improving institutional performance in agriculture and over the long term, enhancing the impacts of these institutions on agricultural productivity. Many national and international agricultural research institutes devote sizeable resources to such activities. However, there has been little methodological and practical work in the area of economic impact assessment of these non-research products.

The recent surge in interest in evaluating the "institutional impacts" of CGIAR centers, has led to significant advances in the methods and application of "institutional impact assessment". This includes the impact an agricultural research organisation has on capacity building, human resources development, and performance of other institutions.

MEASURING WIDER ECONOMIC IMPACTS

Incorporating Spillovers and Spillins

Research often generates four classes of spillovers: (a) interlocational (direct spillovers), (b) interfoci (indirect spillovers), (c) intercommodity, and (d) intersectoral. The most relevant for *ex post* impact assessment are the interlocational spillovers, which affect the production technology and prices in non-mandated regions of a research program. Spillover effects are usually measured in terms of the impacts the local technology has on the K-factor of other regions. Interlocational spillover effects may be international or arise within a country between regions when a research program involves multiple institutions or multiple sites within institutions. Spillovers from one research program are spillins to another. Interlocational spillins are not conventionally part of ex post impact studies as they are conceptually and methodologically difficult to measure. However, recently there have been a few attempts at measuring research spillins of international crop improvement research.

In order to minimise the potential for double counting, and in acknowledgement of the sitespecific nature of many research results, Wood and Pardey suggest assuming no within-country spillovers between those locales where a national research program is simultaneously developing new technologies.

But contrary to the perception that biological technology is location specific, spillovers, both direct and indirect, are pervasive within a country and internationally. The analyst therefore needs to carefully identify the characteristics and sources of technology to estimate returns to a specific research program. The study by Traxler and Byerlee show for the case of wheat research in India that simply analysing productivity gains in the mandate area of the program without tracing the origin of the technology can lead to highly biased estimates of returns to individual programs, with returns overestimated for spillin areas, and overestimated for spillover-producing programs.

With spillins, it is more meaningful to estimate the marginal value added of additional investments in research, rather than the average IRR which might high even for inefficient programs. Maredia and Byerlee show that incorporating research spillins in an ex ante assessment of research investments of a crop improvement research program implies that many wheat research programs, especially those for small mandate areas, are inefficient in terms of overinvesting in local technology development rather than testing and releasing technologies from other programs. Thus when evidence of research spillovers and spillins exists, good practice for ex post impact assessment is to conduct a spatially disaggregated analysis which explicitly treats spillovers and spillins.

The information requirements for such highly disaggregated studies are great, and simplifying assumptions are inevitable. Integrating spillins of research in a conventional ex post impact study adds to the complexity of the model, especially if marginal rather than average RORs are to be estimated, and it may not be practical for all program assessments to include them. Many studies ignore research spillovers also on the ground that for the purposes of accountability to funders, spillovers add little value to the analysis. However, if ex post impact assessment is to be used as an input in making strategic investment decisions and research priority setting, estimations of research spillovers and especially spillins is valuable and should be incorporated into impact analysis.

Equilibrium Effects of Research

Agricultural research generates several indirect economic effects, including labor-market effects and growth linkages of technical change, that go beyond the standard measures of consumer and producer surpluses discussed above. The effect of technology adoption on demand for labor has potentially major indirect benefits. These effects will also often operate inter-regionally to the extent that inter-regional labor markets exist. Total effects depend on labor supply and demand characteristics as well as elasticities of supply and demand for targeted agricultural products but are often limited by the low observed elasticity of total agricultural production.

Overall effects of technical change as a result of research ultimately depend on general growth and income effects induced throughout the economy by consumption, input and output linkages. A full accounting of these effects requires a general equilibrium (GE) model. The few studies that have tried to estimate these effects indicate that in many cases, indirect growth-linkage effects rival or even exceed the effects of conventionally measured economic surplus. These studies have estimated multiplier effects of direct benefits of research of the order of 1.3-2.0. Growth linkages are higher where agricultural income is a high proportion of household income, where initial asset distribution is relatively equitable, and where there is underutilised economic capacity.

It is not conceptually difficult to formulate such models within a GE framework but their practical application will necessarily be limited by the information, time and analytical capacity required. The incorporation of the various effects may require regional disaggregation of product and factor markets and inter-regional trade flows.

For large countries which account for a significant share of international trade in the commodity of interest, such analysis should extend to international markets. However, economists have not yet developed GE models that are both practical and detailed enough to provide much guidance for evaluating research impacts. The modeling of the entire economy (or beyond) is an exercise that is likely to be beyond the scope of most research evaluation and priority-setting studies, and the capacity of most AROs to implement. In addition, there is no evidence that including indirect effects of research would lead to a difference in ranking of research program impacts, or in returns to research relative to other public investments.

Impacts on Income Distribution and Poverty Alleviation

The economic surplus approach can be extended to assess impacts on the distribution of income: (a) between producers and consumers, (b) among different groups of consumers and producers, and (c) between producers in different regions. With the emphasis on poverty alleviation as a central objective of many donors and governments, tracing the impacts of research on the distribution of producer and consumer surplus between different income groups is a logical extension of the economic surplus approach. On the consumer side, this is relatively straightforward since consumer expenditure information stratified by income or expenditure group is widely available and benefits can be partitioned between income groups according to their share in consumption. The overall effects, however, depend on substitutions among commodities, requiring a full specification of demand functions.

Producer surplus can also be partitioned to income groups according to their share of production or proportion of production marketed. However, lack of data on sources of producer income by commodity for different income strata is a major constraint to this analysis. However, the increasing availability of household survey data for analysing poverty should help in this regard. More commonly, some proxy, such as farm size is employed.

Market-price effects of technological change in one region also have important price spillovers for producers in other regions, often with negative equity effects. The classic study of Scobie and Posada indicates how increases in productivity of relatively well-endowed irrigated rice farmers produced negative benefits for upland rice farmers who could not adopt the technology, but many other scenarios are possible. With spatial mapping of poverty variables, an interregional equity dimension can also be incorporated.

These studies indicate that the poverty incidence of research can be readily incorporated in extensions of the economic surplus models. More refined measures of poverty impacts can be employed by considering effects not only on the number of poor but also on the severity of poverty. However, application of these approaches is constrained by the availability of data on poverty incidence, especially on the producer side. Also indirect effects on poverty through labor markets and general growth linkages are not usually considered, although there are examples of the application of GE models to analyse broader income-distributional impacts of technical change.

In sum, good practice is beginning to emerge in this area. Given the strong poverty alleviation focus on the CGIAR, IARCs should be leaders in mainstreaming this type of analysis.

INSTITUTIONALISING ECONOMIC IMPACT ASSESSMENT

Impact assessment can be carried out by analysts within the organisation or by independent professional evaluators (including those working under contract to the organisation). Impact assessments designed to influence internal decision making are clearly better done internally. If the objective is to provide information to those who fund the research, the tradeoff must be assessed between the credibility that may accompany "independent" external assessments and the cost-effectiveness of more informed and targeted internal assessments.

Most AROs, including IARCs, should have "in-house" social science capacity to conduct impact assessment. A few IARCs have formalised this capacity into specialised impact assessment units that report directly to the director general in order to provide independence. External evaluators may be contracted by the ARO itself, or by the research-funding agency. Many impact assessment studies have also been done "independently" by university researchers. Results of these academically based studies may provide useful information but their lack of formal connection to decision makers in AROs or funding bodies limits their influence.

Whether evaluation is done internally or externally, the "ownership" of impact studies should remain with the ARO whose research programs are being assessed. The involvement of decision makers and researchers of the research institutes is the key to the successful use of the results of impact studies. Not only social scientists but technical scientists should be full participants in impact assessment in order to bring the best information to the analysis and also facilitate timely feedback of results to research program priorities.

A research organisation should at least have a champion who provides leadership in the area of impact assessment and helps integrate the results of such studies into research planning. The impact of "impact assessment" will depend on mechanisms developed within a research system by which results are communicated and considered as input into the decision-making process. This will depend on several factors:

— making researchers focus on the practical application of their work as well as its technical merit,
— increasing the interaction between scientists and evaluators at the initiation and development stages of program proposals,
— requiring that all new proposals beyond a threshold investment be accompanied by an ex ante impact assessment, and
— making monitoring and evaluation (M&E) of ongoing research a routine practice.

Byerlee and Alex provide some guidelines for institutionalising impact assessment in national research institutes, through the establishment of M&E capacities. Good practice is that once every five years, each applied research program should undergo a comprehensive external review of technical quality, and an analysis of its economic impacts.

The major problem in M&E systems for research has been the tendency to collect too much information in a highly centralised and bureaucratic manner with little thought about how such information can be incorporated into decision making. To date, there are few examples of good practice in this area but the key elements of successful institutional capacity in impact assessment appear to be:

— Designing a simple data system that collects data on critical variables on a regular basis (see below).
— Establishing a very small unit to *promote* monitoring and evaluation, provide training, develop standards, and arrange evaluation of research programs.
— Ensuring that there are mechanisms for research managers to receive timely information from impact evaluations, and
— Contracting out some M&E activities, especially impact studies, to independent agencies or individuals.

Institutionalising Data Systems for Economic Impact Analysis

Impact assessment is a data-intensive activity that demands time series industry-level data (size and productivity measures such as production, yield, area, commodity prices, and elasticities), measures of research outcomes (productivity change, technology adoption, change in cost of production, institutional development) and the research program inputs and outputs that lead to these changes. Collecting appropriate data is perhaps the most timeconsuming and costly component of conducting impact assessment.

Research programs must balance data needs between the ideal (and costly) and the practical, and must draw data from various sources. If impact assessment is to become an integral part of the research process, it is important that an appropriate data system be institutionalised within the research system. Institutionalising a data system also ensures that the information and data generated by research are available in a systematic and timely manner and are retained for future use as the staff and institution change and evolve. Data generated internally by a research organisation include experimental data (both off-farm and on-farm), data from adoption surveys, data from informal surveys and case studies.

Management information systems such as the International Crop Information System are increasingly being employed to store and retrieve experimental data in the IARCs and some NARSs. However, comparable data systems for adoption and farm survey data have yet to be developed, and have normally been implemented in an ad hoc manner in each ARO. A key issue for many AROs is the role of baseline and panel surveys to provide data on adoption and related household variables. Many types of impacts can only be adequately assessed if relevant baseline data exist (especially for disembodied research products), and regular re-surveys, preferably of the same households, are undertaken over time to monitor changes in farmer practices. However, the resources required for comprehensive panel data sets are usually beyond the capacity of most AROs. National research organisations for example, cannot conduct national baseline surveys both because of the large sample size required for national coverage, and the fact that data requirements for tracking impacts of all research programs (various crops, livestock, crop protection, natural resources management etc) are prohibitive.

Compromise must therefore be the order of the day. We think that each ARO should have a panel of households that it regularly monitors for changes in key farm practices and productivity indicators related to the most important types of research outputs. The sample should be representative of an important subset of the mandate area of the ARO, but few AROs will have the capacity to monitor more than 100-150 households. Monitoring can be undertaken at intervals of three to five years to track changes in practices, and more detailed data may be collected from a sub-sample of households. Increasingly with emphasis on natural resources management, such monitoring should include field observation of soil and water quality, pest populations, and biodiversity, which adds considerably to time and resources required. While such data sets will be far from comprehensive for the purposes of impact assessment, they do provide a base that allows 'informed extrapolation' to a wider domain.

IARCs have been in the forefront in setting up farm-household and farm plot panels. However, even some of these panel data sets have not been sustained over time. In addition, to our knowledge, none of these data sets has yet been geo-referenced using global positioning technology to allow future researchers, and future generations of researchers, to accurately return to the same fields and farms in order to up-date the information.

Such data collection systems necessarily require considerable up-front investment to define the key parameters, sample size, frequency of data collection, and benchmark sites. However, once established, the collection of data at regular intervals has been shown to be a relatively low-cost exercise. Collection of such data should become an integral part of farming systems and natural resource management research programs at major research stations in national systems.

In the meantime, district-level secondary data offer the best opportunity to analyse long-term trends in resource quality. For example, in Pakistan, soil and water test data have been collected continuously for over 30 years in response to demands from farmers and extensionists but only recently have been compiled and made available. Similar data for Indonesia and parts of China have also provided new insights into long-term trends in soil quality. Measures of genetic diversity in farmers' fields have recently been developed, which may be proxies for pest and disease losses. Wider use of simulation models also has potential to infer trends in resource quality.

Good practice in impact assessment suggests increased efforts by IARCs to work with national AROs to establish and maintain relevant farm-household panel data sets for impact assessment.

Presenting the Results of Impact Assessment

Research impact studies generate much information and are sometimes very technical. Research managers and policy makers also tend to be skeptical of the data and methods used; they often also find the reports difficult to understand, interpret, and apply. Therefore presenting this information in an appropriate format and through appropriate channels is important to enhance communication with scientists, research managers and funders. Preparing policy briefs summarising the results of impact studies in non-technical terms is good practice for disseminating the results. Relegating the more detailed explanatory material to the technical appendices of impact reports will provide the plausibility-enhancing material in a place where the specialist can readily find it.

Timeliness is also a key to utilisation of impact information. Often program managers and staff have mixed feelings about impact evaluation—they recognise the need for more and better evaluation, but their personal experience is that evaluations have seldom produced information when needed for decision making.

A key question for research managers is how much to invest in impact assessment. Cost of impact assessment must include:

— Direct costs of evaluation staff, travel, field work, building the data system, data analysis, and overheads.

— Opportunity cost of the time of those involved in the evaluation process which is especially high for social scientists, who are typically in short supply in most AROs, and who must carry out field-based diagnostic studies, farmer participatory work, and ex ante priority setting.

These costs must be weighed against benefits that may be derived from the process itself, as well as the results of the impact assessment. Program scientists often benefit from the rigor of carefully defining the products and beneficiaries of the research program, and estimating model parameters, such as the K-factor. AROs benefit if the impact assessment leads to increased efficiency in resource allocation, or level of funding.

Thus, like any research activity, impact assessment should be subject to the same pressures for justification as other uses of research funds. Some expenditure on research evaluation is surely justified in any system, but there are no clear guidelines on how much to spend. GTZ aims to spend on evaluation up to 5% of a research program's budget on average. INTA in Argentina spends an estimated $2 million on M&E, or 2.5% of its entire budget. One of the summary recommendations of a research evaluation workshop organised in Australia was that 3-4% of the research and development budget could be spent profitably on impact evaluation. The trend towards collecting data and measuring past productivity growth in response to increased demand for impact analysis means that the cost of ex post evaluations may fall. But this may be offset by the trend toward evaluating the social and environmental impacts of research, which are likely to be more difficult and costly.

To institutionalise impact assessment each research institute needs to develop guidelines on budget share for impact assessment, and these costs should be included up front in program budgets. With further developments in assessing impacts of social science research, it may eventually be possible to compute the economic return to impact assessment.

REFERENCES

Akino, M. and Y. Hayami, "Efficiency and equity in public research: Rice breeding in Japan's economic development", *American Journal of Agricultural Economics*, 1975.

Alex, Gary, "Assessing Agricultural Research. Towards Consensus on a Framework for Performance and Impact Assessment", Agricultural Research and Extension Group, *Special Report No.6*, World Bank, Washington, D.C., 1998.

Davis, J.S. and G. Lubulwa, "An overview of ACIAR's economic evaluation activities with an animal science program focus", *Economic Evaluation Unit Working Paper Series No.15*, Canberra, Australia: ACIAR, 1995.

Maredia, M.K. and D. Byerlee, "Efficiency of research investments in the presence of international spillovers: Wheat research in developing countries", *Agricultural Economics*, 2000.

8

Relativism in Horticultural Research and Development

Learning, it is assumed, creates the framework conditions in which the altruism inherent in human nature can manifest itself and bring about social justice. In this, the notion of learning keeps a progressive stance, but it fails to elaborate on the meaning of 'learning'. How is it supposed to advance the interests of the poorest, or address issues of arbitration in their favour? Scientifically, the notion of a participatory or learning 'paradigm' itself is not clear in the discourse of its promoters. The justification ranges from the strictly scientific – drawing on various constructivist theories and cognitive relativism – to the moral and the metaphorical, with a reference to theories drawn from various bodies of science to argue for relativism and participation. The limits between science and metaphor are blurred.

By trying to justify participation and learning by reference to grand philosophical principles which are difficult to argue rationally, the alternative paradigm does nothing but reproduce the contradictions it criticises at a higher level. Basically, its objectives remain the same: modernisation through the transfer of models of social processes.

POST-MODERNIST THINKING IN DEVELOPMENT STUDIES

Post-modernism, as many authors note, can be conceived of either as a critique of modernism or as an attempt to transcend it, depending on whether its stated aims are rejected or considered as already accomplished and in need of review. Perhaps the simplest way to define its philosophy is to say that it is a questioning, if not a rejection, of the values of the Enlightenment and its attendant belief in the power of positivist science to uncover the laws of nature, promote development and build a better world. Within this post-positivist, broad-based philosophy, there is such a diversity of conceptualisations that there is some doubt as to whether the term 'postmodernism' refers to anything of any real substance.

'There are as many post-modernisms as there are post-moderns', 'post-modernism is but a cacophony of voices' are two often cited expressions. Yet the term, despite its imprecision, cannot be easily dismissed, because it encapsulates a feeling that a threshold is being crossed in relativism with little sense of direction beyond it. This feeling is perhaps stronger in the field of development, where the sense of nature and direction of change has always been central to dominant theories.

Development, it is said, is at an 'impasse'. The fading belief in the ability of a scientifically conceived productivist model to solve the problems of the Third World leads some to question the source of this belief, that is, the ideology of progress at the root of modernity. How to refer to such a broad spectrum of approaches without falling into the bias of reducing them to their lowest common denominator? How to avoid the confusion between positions that are historically and theoretically far apart?

Historically, the technological and industrial achievements of the post-war era marked the rise of modernist theories while the social, economic and cultural crises of the seventies led to disillusionment with modernising policies, even if scientific achievements still went unabated. The notion of development, as the embodiment of the belief in an objective of modernisation which marked much of the post-war era, is itself the object of a strong critique. This critique, closely linked to the 'impasse' or 'crisis' in development studies since the late seventies comes, according to Booth, from three perspectives. A first is the post-marxist, or neo-Weberian, rediscovery of social, cultural, sexual diversity and their specific patterns of domination.

While breaking away from the dogmatism of marxist categories, this influence remains to a large extent marked by structuralism. The second is the strong influence of the 'constructivist' perspective, drawing from the tradition of social phenomenology and interactionist anthropology. The third influence, somewhat in-between the first two, is the 'actor-oriented' type of sociology which attempts to bridge the gap between interactionism and structuralism and provide a theoretical solution to the old actor versus structure dilemma.

One could say that the three relativist perspectives, while offering a critique of modernisation and dependency-type theories, attempt to go beyond the 'impasse' in which development found itself through an emphasis on the multiplicity of interpretations, arguing for subjectivism and emphasizing micro-sociological analysis. In this continuum of relativism, based on a critique of positivism and of macro-theory, post-modernist theorists tend to occupy the extreme end.

As Barnhart notes, in the view of post-modernists, 'any attempts to construct a "grand theory", necessarily dismiss the naturally existing "chaos and disorder" of the universe, placing it in an arbitrary framework which, in the words of Michel Foucault, fails to "respect..differences"'. Thus, against the central modernist belief in a universal system of knowledge and values around which practice should revolve, post-modernists argue that such a system rests on foundations that are socially and culturally constructed and that any meaningful analysis should, first, question the validity of those constructions. Such view conveys, a contrario, a set of assumptions about society, about the nature of politics and the role of intellectuals and the nature of enquiry. These have been summarised in Table 1 below.

Modernisation is seen as bringing order through integration of individuals and communities by processes of Weberian bureaucratisation, industrialisation and markets. The contradictory dynamics between order, brought about by modernisation, and freedom, inherent to human nature, is therefore highlighted by post-modernist thinkers, whose analysis suggests counter-models of social organisation and political action founded on different social categories and aiming at different objectives.

As Domingues points out, 'the disembeddings provoked by modernity of individuals and collectivities from contexts which provide solid identities calls for an unceasing creative activity, so that new identities substitute for those ruined ones'. These new identities are rooted in the 'capillary, localised and decentred forms of resistance' to the 'modernising offensives' of the élites. These forms of resistance are what Escobar and others call 'New Social Movements', that is, movements whose foundation escapes the usual categories of social classes.

Escobar notes that 'one cannot look at the bright side of modernity without looking at the dark side of domination. Modernity's inventions must be seen for what they are: Janus faced relations between forms of knowledge made possible by reason and the systems of power created in the process of building a rational society, including the processes of emancipation'.

Although devoid of the strong critique of 'normal science', a similarly discursive view of development theories, whether mainstream or 'alternative', is shared by actor-oriented sociology which attempts to conceptualise the dynamics of power relations at the interface between development institutions and local populations, Arce. Long, one of its prominent advocates, argues that development theories are fundamentally biased by determinist and centralist thinking, assuming social change to occur solely through external intervention and ignoring local actors' room for manœuvre. Thus they are 'tainted with a dreadful sense of fatalism' and a serious under-estimation of the agency of actors who 'try to create space for their own projects'.

Table 1. Assumptions of modernism and post-modernism compared

Modernism	*Post-modernism*
Human Nature and Society	
Order/Stability/Determinism	Chaos/Anarchy/Unpredictability
Subordination	Freedom
Dependency	Autonomy/human agency
Politics	
Development as an objective of progress	Development as a discourse
Power lies with structures	Power is defined through relations/ language/reification strategies
Classes, parties	'New Social Movements'
Rooted in economy/sociology	Rooted in identities
Politics is a distinct sphere	Politics lies in the social sphere
The political transcends the personal	The personal is political
Role of Intellectuals/Practitoners	
Provide 'narratives'/Create economy of truth/Legitimise power	Deconstruct 'narratives', sort out the relationship knowledge/power
Provide analyses/solutions	Facilitate communication
Nature of Enquiry	
Science ('meta-narratives')	Literature ('deconstruction')
Universal/ Deductive	Local -Specific/Inductive

Monologic discourse	Dialogic discourse
Reason	Passion
Certitude/ Determinism	Ambiguity/Indeterminacy
Precision/Disaggregation	Allusion/Holism
Announcing	Questioning
Re-affirmation	Subversion
Continuities/Constants/Unity	Discontinuities/Singularities/ Diversity
Convergence/Simplification/ Synthesis	Divergence/Complexity/ Difference
Exclusion of alternatives/margins	Inclusion of alternatives/margins
'Facts'/Sharp images	'Constructions'/Broken images

In so far as post-modernism is associated with a critique of modernisation as a theory (if not of modernity itself as a concept) and, through it, with issues of language and power, actor-oriented sociology is very close to post-modern theory, albeit of the type Rosenau calls 'affirmative', that is, a relativist stance that remains optimistic and turned towards progress. Like post-modernism, actororiented sociology attempts to decrypt language, treating development as a discourse and using the deconstructionist technique of literary criticism, to demystify power in a Foucauldian model of analysis.

Development as a Discourse, Knowledge as Power, Science as a Construction

Science is a central target of the 'deconstructive' drive in development. The criticisms levelled at it come from different perspectives but they share a common argument: 'the mental productions we call scientific knowledge are no less subject to social influences than the products of any other way of knowing', and therefore they construct more than they describe the 'scientific truth'. Science, it is argued, plays a prominent political part in the modernising efforts. It does so through logical positivism, its dominant paradigm.

Historically the development of the scientific method is closely linked to the ideology of the Enlightenment of which it is both the driving force and the legitimation. More specifically, because it is linked to the notion of progress and objectivity, science has forged an alliance with the state in a progressive movement, against the church. As Norgaard sums it up, 'the progressive movement..had politically neutral scientists making decisions on the public's behalf under the broad guidance of elected officials'.

In development, as an emanation of western civilisation, science's understanding of the world is bound to western culture and values which it attempts to diffuse. To Escobar, 'the production and circulation of discourses is an integral component of the exercise of power..(and) development itself, as a discourse, has fulfilled this role admirably'. It has done so, according to Escobar, through a twin process of professionalisation and institutionalisation. The first is the creation of techniques and disciplinary practices that act as a mechanism for the creation and maintenance of an 'economy of truth' and the 'production of norms' which, in effect, act as a tool for the control of the Third World.

The arsenal of disciplines and sub-disciplines spawned by the discourse of development produces specialised knowledge and legitimation at the same time as it socialises professionals into a model of thinking about the Third World. Institutionalisation, through aid agencies, universities, international organisations, voluntary agencies etc., in effect, acts as 'a network that organises visibility and makes the exercise of power possible'.

Such an apparatus, which 'is a significant feature of the rise and consolidation of the modern state..inevitably structures the encounter of the organisation and its "clients" (e.g. peasants) in such a way that the latter's local reality is transcended and elaborated upon by the former'.

The impact of this encounter is all the more pervasive since what is being imparted on the Third World is not simply a capitalist system of production but also a system of power and a system of signification. Thus, the effect of the development discourse 'has to be seen not only in terms of its social and economic impact but also, and perhaps more importantly, in relation to the cultural meanings and practices they upset'.

This upset of practices is tantamount to a destruction of local knowledge and initiative and, concurrently, a growing dependence of local people on development experts. Thus the development enterprise, rather than alleviating poverty and empowering the masses, has modernised poverty and empowered an élite army of development professionals. Beyond the discursive turn, the root of this process, it is argued, lies in the built-in tendency of positivist science to 'produce ignorance' and destroy alternative interpretations. As Arce puts it, 'a body of knowledge is constructive and creative in the sense that it is the result of a great number of decisions and selective incorporations of ideas, beliefs and images, but at the same time destructive of other possible frames of conceptualisation and understanding', including the 'desecration of local knowledge'.

In this sense, science could be said to decrease our knowledge, not increase it, because the belief in universal theories prevents the exploration of other directions and frames of interpretation. The main focus of the critique of positivist science in development is undoubtedly the field of neo-classical economics. Professionally, this focus is justified by the influence that economists have on policy making, as compared to other social sciences. Scientifically, economics is seen as the embodiment of an imperialistic methodology attempting to bend reality to a set of universal 'laws' patterned on classical physics.

In economics, it is argued, concepts such as competition, market information, supply and demand etc., are presented as 'objective' facts, built in abstract mathematical models that act as a screen hiding power relations and value-laden judgements. These models protect economists from the messy nature of real-world data and represent a 'constructed reality' whose main features are reductionism and measurability. Methodologically, economics reduces human nature to a Pavlovian reflex, responding solely to economic incentives Chambers and the complexity of human motivation to the notion of a utility-maximising 'homo economicus', excluding social, ethical and spiritual issues. Policies designed within this materialist conceptual frame are bound to favour the economic status quo through an emphasis on (an already unequal) market solution, encourage the acquisition of material goods and undermine communities' other values.

The implication of these wide-ranging critiques, in terms of action, is tantamount to an abandonment of the development 'paradigm'. The language of 'participation' and 'empowerment' must be recognised for what they are, i.e., alternative 'inventions' of the development discourse, and Escobar warns us against attempts 'from dominant forces to salvage development through fashionable notions such as "sustainable development", "grassroots development", "women and development", "market-friendly development"'. 'Participatory' develop-ment in a post-modernist view is clearly one such 'salvage' attempt, and a variation around the same discourse. Rather than alternative development, one should speak of an alternative to development.

Science in the time of chaos and uncertainty

Chaos theory, with the sense it conveys of anarchy, randomness and personal freedom, seems to have captured the spirit of relativism. Philosophically, chaos challenges the very foundation of western modernist ideology based on the notion of creating order and controlling anarchy, that is literally avoiding chaos. It is therefore no surprise that it is referred to in post-modernism, and increasingly in the social sciences, either as metaphor or as an attempt to unite all science – social, biological, physical..into a grand, meta-theory ruled by the same general philosophical principle of nondeterminism. Thus in a relativist stance, reference is often made to chaos theory, complexity theory, non-linear systems, cybernetic etc. Trying to clarify these overlapping disciplinary fields is not easy. The difference between chaos theory and complexity theory lies in the emphasis on undeterminism and unpredictability.

Cybernetics is itself an umbrella for a variety of disciplines ranging from systems theory to information theory, even chaos theory, and thus provides a good illustration of the attempts to draw a bridge between physical and social sciences in the form of 'socio-cybernetics'. First-order (or classical) cybernetics has an engineering approach and is concerned with the tracking of negative feedback loops impairing system performance, to design control mechanisms necessary to steer the system towards stability, assumed to be its 'normal' state (homeostasis). Secondorder cybernetics has a strong biological basis and includes the observer in the systems being studied. It emphasizes understanding of the system dynamics of change rather than control of system stability. Its focus is on the positive feedback loops behind the dynamics of change, assumed to be an inherent feature of systems (morphogenesis).

While the epistemological basis of first-order cybernetics rests in a Newtonian philosophy associated with positivist science, second-order cybernetics is clearly constructivist, sharing the notion of knowledge being actively produced by the subjects, including the observers, and not a property of some outside reality. While first-order cybernetics, and beyond it, first-order (or normal) science, attempt to arrive at an understanding (or model) of reality that enables control over it, second-order science emphasizes the unpredictability of outcomes in intervention and a general principle of self-organisation.

Smith notes that 'at the intuitive level, at any rate, chaos theory seems to provide a means of escaping structuralism's two significant flaws: it is inherently dynamic and time-sensitive and it permits a definition of social structural entities in such a way that if real, living, unique human beings vanished then the structures of society would also vanish'. Indeed if chaos theory stands for the large consequences that small changes in initial conditions can have, the theoretical support it can lend to the sociological notion of human agency and the refusal of fixed and deterministic structures is obvious: human beings matter; their impact at a macro-level and their long term effects can be considerable, albeit unpredictable.

If, as Geyer reports, computer simulation experiments show that autonomous systems 'give meaning to their interactions on the basis of their own history rather than on the basis of the intentions of the programmer', this shows system emergent properties not unlike the notion of agency being an inherent feature of human communities, where action – not communication between actors – leads up to outcomes unpredicted by outside intervention.

This view of self-steering, self-organising systems, borrowed by post-modernists from complex systems theory, has relevance for the critique of development theories, particularly in the notion of development as planned change. Here, there are strong parallels between the importance of the non-determinism of second-order cybernetics and the deconstruction of development planning of actor-oriented sociology. From a socio-cybernetics perspective, Geyer argues that system steering implies social change over a time period during which uncertainty increases, conflicts erupt and planning preferences may change, making planning perhaps interesting but inefficient in the first place.

Besides, 'perfect planning would imply perfect knowledge of the future, which in turn would imply a totally deterministic universe in which planning would not make a difference'. Arce echoes from a sociological perspective that social planners believe 'that planned intervention and policy models are able to predict and control the future…political authority is based upon a sequence of steps by which "science" and "technology" impinge on society, a faith that knowledge is objective and neutral, and a sort of discourse that underlines the image that development is concerned with technical considerations alone….science and technology are seen as universal and external artifacts equated with a "modern notion of truth"'.

For this reason, development planning should abandon prescriptive, goal-oriented decision making and prediction about future states and focus instead on understanding the dynamics of change and promoting a collective learning framework through which concerned stakeholders can constantly, through dialogue, express their respective interests and reach consensus.

DECONSTRUCTING AGRICULTURAL RESEARCH

Not all critique of concepts is a deconstruction of discourse and one should be wary of not imparting a grid of analysis through the artificial adoption of post-modernist jargon and thus submit forcefully various approaches to a post-modernist interpretation. Such a question is important when one attempts to explore the influence of the post-modernist movement on the participation movement in agricultural research. Both are sufficiently diverse, scientifically, politically and methodologically to allow any comparison and any contradiction. However, research approaches and methodologies being closely related to broader development theories, the emergence of post-modernist analyses in the development field can be expected to have an impact on them.

As Coffey et al. remark, field research has become fragmented under the influence of contemporary cultural perspectives, such as poststructuralism, post-modernism, feminism and post-colonialism, through which 'runs a discursive turn, treating as central but problematic the relations of language, knowledge and power'. In agricultural research, some of these analyses refer explicitly and identify with post-modernism. In others, the overlap with post-modernist thinking is very clear, not only in the references to the same theoretical sources but also to the same type of argumentation. The central issue is their relativism and the extent to which it can be equated with post-modern theory and this is a slippery issue since, as Reiter points out, 'many post-modernists do not go to relativist extremes and it may not seem very radical to admit that all views of reality are necessarily partial. But this is a post-modern notion'.

The need to deconstruct agricultural research and reconstruct it along different lines is expressed by Kloppenburg in these terms: given all the ills created by normal science-based agriculture, can this same science promote alternative forms of agriculture? In other words, can the cause of the problem be the source of the solution? The answer is clearly no.

Post-modern agricultural research is a thematically and methodologically fragmented model. It must necessarily integrate what has been excluded by positivist science, internalise externalities, contextualise rationality and this requires that it be socially reconstructed with the inclusion of the knowledge production capabilities of all marginalised groups. Consistent with the critique of positivistic hegemonism, it must be methodologically both pluralistic and holistic; this is the condition for it to be thematically gendered, communautarised, localised and ecologised, that is to integrate the social and environmental perspectives of multiple actors. The rationale for each of its elements is briefly described.

Holism and Reductionism

According to post-modernists, science was never meant to understand the fundamental nature of human existence but to find out regular and recurrent patterns in nature so as to act on them and achieve stability, regularity and predictability. As Norgaard says 'western science has sought to know the universal, unchanging characteristics "behind" a changing reality' and to provide 'one consistent set of laws about the nature of all things'.This search for regularity is done, methodologically, through reductionism and aims, according to Kloppenburg, to produce through abstraction and simplification, what Latour calls 'immutable mobiles', that is constants across social and spatial locations.

Thus it ignores the political, physical and biological specificities of the local context, one of which is the 'intimacy between the worker and the materials and objects of labour'. It is this intimacy which makes the unique character of the 'mutable immobiles' of local knowledge, easily marginalised by normal science's attempts to apply universal laws to it. So pervasive is reductionism, it is said, that from a method, it becomes a way of looking at things and its main approach is to 'problematise' them, that is to turn them into problems to be 'fixed', thus denying the potential for self-improvement.

Problem-orientation and wish to control are similar, in cybernetics terms, to the aim of correcting negative feedback loops, that is the notion of 'command and control' in management. They have implications for methodology. As Meppem and Gill say, they largely ignore non-linear positive feedback, that is the potential for destabilisation or, on the contrary, of growth of the system. Their cost is 'an unnecessary constraint on the adaptive ability and creativity of an organisation or economy' . In terms of research, Geyer specifies that 'if one investigates a certain system with a research methodology based on the control paradigm, the results are necessarily of a conservative nature; changes of the system as such are almost prevented by definition..A different methodological paradigm is needed if one wants to support social change of a fundamental nature and wants to prevent "post-solution" problems; such a paradigm is based on a multiple-actor design, does not strive towards isolation of the phenomena to be studied, and likewise does not demand a separation between a valuedependent and a value-independent part of the research outcomes'.

In the PRA movement, the impact of reductionism in agricultural research, intimately associated with normal science, is described by Chambers. In project planning, the Logical Framework approach, for example, reduces the complexity of development to a single central problem identified mostly by outsiders. Thinking about poverty is reduced to a narrow technical definition meant to ensure that a scientifically universal understanding and a unique standard of measurement are achieved. Agricultural production is reduced to yields per hectare. Food is reduced to grain production.

Employment is reduced to concepts derived from the formal urban sector in the industrialised north. Chambers comments that 'in reductionist thinking, in a mysterious way, intelligence and humane common sense seem inversely related. And since intelligent people dominate the discourse of development, we can expect reductionism to remain robustly sustainable'.

In the context of agricultural experimentation Chambers and Jiggins explain the causes of reductionism's bias against resource-poor farmers. They point out that reductionism excels 'in exploring the relationships of a restricted number of variables in controlled conditions. This suits it to a large-scale simplified farming in which the natural environment is highly controlled, with monocropping and standardised mechanical, fertiliser and pesticide treatments'. This model of agriculture is mainly that of resource-rich farmers who are financially powerful and politically organised enough to influence research policies. In addition, they are culturally close to researchers in terms of values and attitudes and are able to communicate more easily with them.

This is contrasted with the conditions under which resource-poor farmers operate. They have smaller farms and different priorities, their physical environment is more diversified and less controllable, interactions are more complex and farmers' management, developed through constant adaptation is of paramount importance. They need a research that is locally-specific, and therefore flexible, whereas reductionist, 'normal' science emphasises genotypes and environmental conditions at the expense of management and interactions. 'For reasons, thus, which..are environmental, political, social and methodological, most agricultural science has a bad record in serving farm families who are resource-poor'.

Reductionist science is at the heart of the 'teaching' model of transfer of technology and, in general, of the diffusion of innovation model which has always dominated science. In short, as Chambers and Jiggins say, this model, embedded in normal science, is traditionally oriented towards products and not towards clients. For this reason, they propose a 'farmer first and last' model where, against the notion of transfer, everything starts and ends with the farmer.

In short, to redirect science towards the 'client' and to allow the 'mutable immobiles' of local knowledge to emerge, the research process should therefore be participatory (including all stakeholders), open-ended and holistic; its structure should not be pre-determined by hypotheses and its evaluation should be based on different criteria. However, holism here must not be associated with researchers' system thinking. Systems as drawn by researchers carry with them assumptions and limitations.

The scientists' systemic thinking, it is argued, is dominated by the hypothesis according to which the system has an intangible reality: it is therefore simply a question of understanding its components and functioning and identifying its problems and solving them. 'In this view, the natural and the social world are essentially seen as predictable, controllable and therefore optimisable towards previously defined goal..(that is)... a type of passive conceptualisation of human action'. This hypothesis predetermines the procedures of problem analysis and resolution, including the type of diagnostic survey and the level of farmers' involvement. Thus Chambers and Jiggins insist on the shortcomings of Farming Systems Research, be it its costs, its method or its philosophy. Kloppenburg also indicates that the solution to problems at the whole farm and local system levels does not lie with agricultural researchers but with 'those who think in terms of whole farms, those whose experiences are of whole farms, and whose knowledge has been developed by the integration of hands, brain and heart in caring labor on whole farms – that is..farmers'.

Communitarisation and Localisation

The emphasis on communities and local-specificity in post-modernism is a natural consequence of its rejection of universal models. Whether from the point of view of analysis or of action, a locally and socially 'situated' approach is central. In actor-oriented sociology, in post-modernism and in constructivism, empowerment lies in human agency and this agency is socially and culturally specific. Thus Kelly and Armstrong note the 'need for a more situation-specific and place-sensitive consideration of the power relationships which underlie poverty and inequality in the developing world'. Meppem and Gill, more specifically, point out that conflicts are value-ridden. Values are not universal and cannot be applied abstractly and mechanically, they must be considered, and applied, in their relevant context.

In agricultural research, Pretty and Chambers add that 'the dominant positivist framework has missed local complexity; determinist causality has failed to account for the adaptive performances of farmers; technologies successful in one context have been applied irrespective of context, with widespread failure...'. From a learning perspective, the local context is of a central importance. The notion of 'situated learning' is inherent in constructivism: knowledge is created and made meaningful by the context in which it is acquired, and learning is embedded in the daily experiences of the learners. Methodologically, social and locally-specific research is the only way to re-integrate the 'mutable immobiles' excluded by the reductionism of normal science.

As for the communities' relation to the wider economic and political contexts, their weakness is also a source of strength because, as Perkins writes, 'it is exactly the fact that communities are somewhat removed from national and international "public life" that can make them strong bulwarks against centralised control, refuges of diversity, and incubators for creative human interaction'. Households, as intermediates between the personal and the community, and communities as intermediates between the 'public' and the 'private' sphere must be the primary focus of analysis and action.

Ecologisation

For political and methodological reasons, positivist science, it is said, cannot deal with the environment. Methodologically, reductionism means that 'science fails the environment'. While the environment is characterised by multiple causations, science deals, through reductionism, with broken down parts that allow precision, establish clear cause-effect relations and draw general laws, that is produce certainties. Normal science is unable to deal with uncertainties, when uncertainty is precisely what characterises the complexity of ecological processes. This 'uncertainty comes with even the best available science because natural resource management questions are fundamentally ambiguous'.

Controlled experiments in this context are difficult and results of investigation are rarely unequivocal. Economics deals with monetary values when the environment is precisely difficult to value and economics itself is not a unified theory. Thus traditional valuation methods – mostly cost-benefit analysis – impose a rigid system favouring market-derived values that are amenable to quantification, to the detriment of intangibles such as aesthetics or impact on cultural integrity. These failures have strategic, methodological and political implications.

In the past, science could use the uncertainties of the data as lack of proof and therefore the burden of proving a threat was on environmentalists. For science to accept uncertainty as an inherent feature of complex natural systems and incorporate it as a philosophical principle, it needs to recognise the possible long term threats to the environment. This means that strategically it should adopt a 'precautionary principle' and take on itself the onus of the proof.

Methodologically, as far as the environment is concerned, 'good science' is necessarily holistic. 'It is self-defeating and counter-productive to continue the pursuit of more solutions within the traditionally isolated intellectual spheres of economics, politics, technology or education, because the causes of the global (environmental) crisis are no longer purely economic, technical, social, etc.'. Being holistic means also reintegrating in science the cultural dimension that is inherent in the relation of man to his environment in order to balance out the economic bias. Science should also abandon prediction and control in favour of explanation and learning. Complexity and uncertainty mean that science will never be able to have all the answers and that 'no single party, agency, organisation or discipline holds the key to understanding a particular natural resource management situation'.

What is therefore required is the largest possible consultation, built into a permanent process of social learning. Recognising the necessity of social learning is, in fact, adapting to two specificities of the relation between science, politics and the environment. First, there can be no distinction between science and politics in the management of ecosystems: conflict is embedded into the management planning process. Second, scientific and political systems operate on different time scales. 'The complexity of ecosystems prolongs the datagathering process, and that in turn, complicates the policy decision-making process'. A permanent feedback is therefore required, in forums that encourage mutual learning, between the community of users, economic and political decision-makers and different areas of science.

Genderisation

A fourth dimension of a post-modern agricultural research has to do with the integration of a feminine perspective. The need to 'genderise' science is not simply a question of taking into account the specific needs of women nor to heed their voices, that is ultimately a question of social justice.

In a post-modern agenda, it is also a question of integrating a feminine worldview and the specific feminine mode of knowledge production, that is a feminist epistemology. The specific feminine worldview is perhaps best illustrated in the field of feminist economics. Perkins, for example, argues that man's work is paid work, it is a time-oriented social time; it has been distanced from the natural rhythm of ecological time and the human physicality with other sensual beings and the natural world.

By contrast, women's unpaid work is task-oriented, biological time, born out of a life-long socialisation into caring and nurturing activities. As such it holds life-affirming values, emphasizing co-operation, empathy and nurture, focusing on process rather than results, valuing intuition, subjectivity, creativity and spontaneity. Feminist theoretical work, therefore, is 'vitally important for understanding communitybased approaches (for its values are) similar to those held by aboriginal cultures and the ecology movement'. For this reason, women play a central role in maintaining community bonds: it is they who are the guardians and transmitters of the values and skills that reproduce communities. A gendered science would, in effect, be both community-enhancing and 'naturally' conducive to a more ecologically friendly research.

Thus is outlined a post-modern agenda for research. Its main elements – communitarisation, localisation, genderisation and ecologisation – are self-reinforcing. Communities are evidently localised, their knowledge is local-specific and more attuned to their ecological environment. Human agency is culturally specific and it is within communities that it translates best into power. The methodological implications of this fragmented agenda are that research must be both holistic and pluralistic.

Theoretical and Methodological Pluralism

To allow the suppressed voices to be heard, scientists must

accept pluralism and fight the hegemony of any single mode of knowledge production. For example, 'confronting ethnocentrism means moving from conventional definitions and universalised theorisations of development towards more pluralistic conceptions of well-being'.

Norgaard, from the point of view of ecological economics, summarises the case for pluralism. As opposed to a single method – or a 'single way of knowing', pluralism is more appropriate for addressing at least four issues. First, it is necessary when it comes to understanding and managing today's complex problems: the interplay between global economies and local interests, social control of environmental systems and use of scarce resources, adaptation and impact of sophisticated technologies. Second, pluralism helps to avoid mistakes by acting as a constant warning against 'brash action' resulting from the possible biases of a single method; it is a reminder of the complexity of social and ecological systems and the difficulty of taking action. Third, it sustains diversity, be it biological or cultural, by counteracting the destructive hegemony of western forms of knowing, technological intervention and social organisation. Fourth, and logically, it promotes participation and democracy. Since 'the use of a single framework, without modification for regional differences, facilitates control from a single centre of analysis (which) disinfranchises or disqualifies the majority, facilitates the tyranny of technocrats, and encourages centralisation...openness to multiple frames of analysis is a prerequisite to democracy and decentralisation'.

CONTOURS OF THE PARTICIPATORY PARADIGM

Do these elements form the basis of a new paradigm for agricultural research? What is a 'paradigm' in the first place and is this a 'participatory' one? In a post-modernist sense, a paradigm is a model within which one set of truths hold or, to use the 'deconstructionist' literary method, it is a 'text' or meta-narrative'. As such, a critique of a paradigm amounts to its 'deconstruction', that is the demonstration of its limits as a cultural construct in explaining a phenomenon or in solving a problem that other paradigms cannot solve. This leads to a first, major difficulty because of its self-justificatory argumentation. As Hammersley and Gomm argue, 'within any community, relativism can be used to justify enforcement of the epistemic paradigm that is deemed appropriate to that community, allowing no scope for internal dissent about fundamentals.

All challenges to the paradigm can be met by the response that this is what we as a community believe', thus providing immunity from dissent and external criticism. But while all contenders for a new paradigm status do not fall neatly into the post-modernist frame, how to challenge this status without loading it with post-modern ideology, especially when relativists share with post-modernists the discursive treatment of science?

This discursive turn is clearly distinguishable in the PRA movement, with increasing emphasis as the movement progressed from Rapid Rural Appraisal (RRA) to Participatory Rural Appraisal (PRA) and, presently, to Participatory Learning and Action (PLA), with greater sophistication and formalisation of its theoretical foundations. It is not necessary to review here the many criticisms of RRA and PRA, suffice to say that the transition from RRA to PRA and on to PLA had two main features, one is the abandonment of the economistic justification, i.e., the argument of cost-effectiveness, and, second, a methodological and philosophical contextualisation, away from positivism to which RRA was still adhering. In this recontextualisation, the most important feature is a new claim to empowerment. This claim draws theoretically from various sources, most notably the tradition of action-research, in fields as diverse as anthropology, organisation sociology, modern management and cognitive psychology. But, as will be argued later, the diversity of theoretical references brings with it political and ideological contradictions into the definition of the new paradigm. While the central concept remains the same, the tradition of actionresearch is itself divided in the nature of its objectives, the scope of its action, its political philosophy.

Selener thus identifies four schools of participatory action-research. The first is that which aims at community development with a revolutionary emphasis, which one can perhaps associate the francophone tradition of animation rurale. The second and third are those based on the work of respectively Kurt Lewin on organisations and John Dewey in education and, finally, participatory research in agriculture. The language of this latter, where Robert Chambers has been an outstanding influence, is itself divided between the message of liberation of the first, the management objectives of the second and the pedagogical aims of the third. And thus the claim to empowerment can take diverse meanings in the proposed paradigm. But, first, what are the outlines of this paradigm and what role is allocated to PRA, and more broadly to participation, in it? The next sections will examine, first, the terms of the critique of positivism in PRA and the role of learning in overcoming the failures of positivist science.

Critique of Positivist Science

The cornerstone of the critique of normal science in the PRA movement is undoubtedly the transfer of technology model (ToT), a starting point in almost all the writings of Chambers in the last decade. In itself, the ToT model is seen as epitomising the central shortcomings of 'hard' positivist science in different fields: the notion of knowledge as an objective reflection of an outside reality (in philosophy) or an individual property to be transferred (in communication and extension), the instructionist approach of behavioural learning theory (in educational psychology) and in general in the reductionist working of scientific research. The philosophical recontextualisation of the movement away from positivism, is therefore based on learning and organisational models that do away with this notion. But the critique of the ToT is not peculiar to the PRA movement. Röling for example, an outstanding critic of positivist agricultural science, sees agricultural innovation as best approached holistically, in terms of 'soft' models of knowledge and information systems.

Like all relativist approaches and methods, the shift to more holistic and participatory models draws significantly from Kuhn's notion of a paradigm shift, that is the advent of a scientific revolution. In short, this shift is nothing less than a revolution because it is assumed to bring about a dramatic change where the change expected from 'normal' science is usually incremental, detrimental to the interests of the poorest and/or destructive of the environment.

Neither the epistemological ambitions, nor the positioning of field practice in relation to normal, positivist, science are new in the movement. Even in the early 1980s, it was pointed out that, since options are determined primarily by politics, RRA like positivist surveys serves simply as validation of political choices specially when the results of the former are pre-determined by those of the latter. Later, Jameson, sees RRA as a potential method of an emergent development paradigm whose main feature is a rejection of 'evolutionary, unilinear, universalistic, positivistic and utilitarian assumptions' which dominated western science since Bacon, Descartes and Darwin.

While PRA, as such, is never explicitly the object of epistemological justifications, the theoretical constructions its proponents have generated around its practice lead logically to those summarised in Table 2, in the fields of participation, sustainability and indigenous knowledge. If these justifications are considered valid, the principles of PRA, as emphasized by most of its practitioners, take on a special meaning. Their epistemological basis rests with the notions of unpredictability of social phenomena, the subjective nature of 'data' and the endemic nature of 'problems'.

The response to unpredictability and the endemism of problems is iteration (i.e., no pre-set methodology, flexibility, continuous and sustained work with communities) and optimal ignorance (i.e., knowing only what is useful, measuring only as precisely as necessary, being appropriately imprecise). Iteration is the principle through which methodology is adapted to specific situations (actors, problem situation, local environment). Optimal ignorance is the methodological recognition that 'for the complex, diverse, dynamic and unpredictable realities of people, farming systems and livelihoods, comparisons and judgement are often more potent and practical than precise measurement'. As such, it can be seen as a response to the reductionism of, in particular, economics.

The subjective nature of 'data' requires triangulation (i.e., focus on diversity rather than characterise complexity; emphasis on relative rather than absolute quantities) and visualisation, as a central principle. Triangulation is required to account for multiple perspectives because 'all views of activity or purpose are heavy with interpretation, bias and prejudice, and this implies that there are multiple possible descriptions of any real-world activity'.

Visualisation has two meanings. In PRA, it relates to the use of visualising tools drawn by rural people themselves, preferably using local materials. It refers to the need to recognise that 'data' is the outcome of subjective judgements: the scientifically 'objective' analysis is based on the scientist's own cultural, disciplinary, political subjective interpretation. The visualisation process ensures transparency and permits open dialogue and therefore deepens and democratises analysis. Through it, local people use local categories, criteria and symbols and thus share in the creation of knowledge and its analysis: 'they are encouraged to explore their own version of their world'.

The process being more important than the outcome, the PRA exercise can be seen as a means of group negotiation and consensus building. More fundamentally, visualisation means 'making visible', that is, translating concepts and phenomena into non-scientific terms to allow for comprehension and dialogue between scientists and farmers. This translation is necessary for meaningful dialogue since their respective modes of knowledge production are different. Scientists' field of perception is extended by artificial amplification tools (microscope, statistical analysis, computing power etc.), enabling them to reach deeper dimensions of observation when the farmers' is limited to the human natural senses. Thus, if one were to use a management concept, visualisation permits a 'retooling' of research whereby transparency puts everybody on an equal footing.

LEARNING IN A NON-DETERMINISTIC WORLD

The reversal of learning is at the centre of the PRA movement's preoccupations. Pretty and Chambers point out that the new paradigm 'enshrines new ways of learning about the world' that form its central concept. Development professionals are seen as burdened by the thinking entrenched in normal science and need to develop new attitudes and behaviour. The emphasis on learning draws from both constructivist theories of education and complex systems theory. The call for a reversal of learning has two inter-related dimensions. The first is epistemological and draws its arguments from complex systems and chaos theories. It refers to various sources, like the uncertainty principle in quantum physics, information theory and cybernetics to justify a nondeterministic conception of the world and legitimate participation to decision making.

Actors operating in complex systems with blurred boundaries, it is argued, need to learn how to learn in a world (not simply the world of development) whose main features are uncertainty and unpredictability, a constantly changing world. And what is development if not change? The consequences of change are all the more unpredictable since information is faster and sources much more diversified than ever before. Thus, Jameson argues that 'rapid change in both villages and the larger physical and social context in which they are embedded has increased uncertainty and reduced predictability. Change is being promoted without any possible certainty as to its ultimate social and environmental outcomes'. This requires a reconsideration of determinism and replace it, within development, by the twin notions of relativity and uncertainty, now recognised in 'hard' sciences like quantum physics.

Professionally, we must 'confront directly the epistemological challenges of intervention in a complex and dynamic world that is characterised by..inculcate in ourselves and in others greater epistemological humility and flexibility'. because, and Chambers echoes the same concern, 'when faced with the complexity, diversity and dynamism of human and local conditions, there is no normal bedrock on which to anchor, and few fixed points. Rather, we need a repertoire of skills for staying afloat, steering, finding our way and avoiding shipwreck on a turbulent and transient flux'. The only response to this situation of uncertainty is to empower local communities so that they can deal with it themselves.

Methodologically, the implication of 'the emerging paradigm is two major lines of defense against unanticipated negative outcomes' of change; on the one hand, 'involving local people in the selection, design, planning, and implementation of programs and projects that will affect them, thus ensuring that local perceptions, attitudes, values, and knowledge are taken into account' and, on the other hand, 'to make continuous and comprehensive feedback, and the capacity to respond to it, integral parts of all development activities'. The challenge of sustainability is precisely to enhance the local 'capacity to adapt in the face of unexpected changes and emerging uncertainties'.

In the same vein, Drinkwater notes that science is 'situated' in time and in space. Scientists are beginning to recognise that, natural systems being mainly in a state of non-equilibrium, the emerging order is fundamentally unpredictable. Their own action on these systems cannot be conceived in a deterministic fashion. On the other hand, their results will, obviously, depend on the particular geographical conditions but also on the social conditions that will determine them. Therefore they can no longer learn 'from the outside', as if they were simple spectators. 'This historical and turbulent view of nature or society, and the embeddedness of the scientist participant within it, implies that neither nature nor society can be understood independently of the other'. Thus the paradigm is one of multiple embeddedness: of nature and society, of order and chaos, of learning and change, of individuals and institutions, of institutions and environment, of personal change and social change, all intertwined in a complex and dynamic totality.

To develop the understanding and behaviour required to adapt to this embededness, the scientist has, first, a lot to unlearn, among others about his/her own perceptions, attitudes and values if social change is to be effective. This leads on to the second, pedagogical, dimension which draws most of its concepts from new learning theories. It aims at creating 'new' scientists who accept a role of facilitators engaged in a learning process, no longer functioning in a consultant mode, scientists who, metaphorically, are ready to 'pass on the stick' to peasants.

Strategically, the recontextualisation of research into a learning framework can thus be seen as part of the 'humbling' of researchers, whose problem solving procedures are recognised as essentially learning procedures and, therefore, whose status becomes one of simple learners, albeit more effective if they change their methods. In this sense a learning agenda becomes essential for the 're-education' of 'conventional' scientists, associated with positivism, to bring them to shift from a controlling to an enabling mode. Their humbling is an essential step towards the liberation of the suppressed knowledge and the empowerment of local communities.

Drawing from different sciences, the epistemological and pedagogical dimensions share the same notion about the role of learning: in complex systems, whether social, physical or informational, in a truly non-deterministic mode, learning has an adaptive function. In both human and non-human systems, cybernetic principles apply: better communication and permanent learning optimise feedback and, through it, counteract entropy and prevent dysfunctions. Learning takes place through self discovery and gradual adaptations in a system where the locus of control, that is power, is dispersed.

CONTRADICTIONS OF A POLYMORPHOUS PARADIGM

The diversity of the PRA movement makes it difficult to give clear cut answers. The justification of participation is underlined by considerations that are not all, strictly speaking, of a scientific nature but which are closely interlinked. The justifications range from the economistic/ instrumental argument – i.e., participation as a condition for the production or adoption of innovations – to the issue of citizenship/democracy and that of limitations of 'normal' science. The latter justification, there are various social and natural science sources which, explicitly or implicitly, suggest a rationale for participation. For example:

— In sociological terms, social and cultural constructivism provides the notion of reality as being socially and culturally constructed and subject to multiple interpretations; participation is the means through which different perceptions are bridged and consensus reached; research as a process functioning in a Habermasian communicative mode facilitates processes of dialogue enabling the emergence of this consensus;

— complex systems theory and cybernetics (delays of information feedback) in which participation – and rapidity – are seen as a means of providing timely feedback and empowering people to respond to unpredictable development processes, and related to this, more specifically;

— chaos theory which suggests in fine that rural communities, among others, are 'dissipative social structures'in a state of non-equilibrium, whose main feature is chaos from which order will eventually emerge, but in directions that are unpredictable. Although not made explicitly, one can assume that, in this perspective, participation, is conceptualised as the source of 'energy' required by the process of order formation. It is a conception of a non-linear, non-deterministic science.

The three concur in their emphasis on self-organisation and the limits of social science, through research and planning, in determining outcomes. This notion highlights a first, major, contradiction between the argument of autonomy – based on the notion of human agency and a critique of development interventionism common to actor-oriented sociology and more generally to postmodernism – and the idea of externally induced participation and/or empowerment (that is, basically another form of development interventionism), in the PRA, and more generally in the participationist movement. The implications of this contradiction are discussed later.

The diversity of philosophical, theoretical and disciplinary sources makes participation almost a stopgap universal remedy, a philosophy 'by default', a 'catch-all' method and a cost-effective technique, meant to make up for all the gaps in all other conceptual frames. However, diversity and pluralism do not come without their cost. That of the paradigm outlined is a polymorphism that allows it to absorb and neutralise criticism through a constant change of registers, at the expense of political efficacy. Despite the various authors' claims, the diversity of theoretical sources brings serious contradictions to the paradigm outlined. Instead of lending it richness, it brings it poverty and ambiguity: of social theory, of political message, of theoretical foundation, of methodological base.

In short, what it gains in theoretical richness, it loses in political efficacy. More exactly, the theoretical radicalism with which normal science is criticised results in methodological contradictions which in turn are solved by theoretical backsliding and the recourse to metaphor. The polymorphism of the paradigm enables the movement to have several discourses at the same time and to use different registers in the field of knowledge. This is based on the constructivist notion of 'reality', (i.e., cognitive respect and in the field of action), which is suggested by a frequent reference to the Freirian tradition, itself founded on the notion of 'consciousness' and alienation, or to notions of modern management techniques. Thus, in a post-modernist stance, this polymorphism is justified through a celebration of methodological pluralism and a supposed neutralisation of contradictions by their absoption as part and parcel of the paradigm.

A first major difficulty in assessing the participation 'paradigm' is the meaning given to it by its proponents. 'Paradigm' is sometimes used in its scientific, Kuhnian, sense with a clear wish to give participation a scientific legitimacy, as distinct from a moral or political legitimacy. Rarely do its proponents refer to fundamental issues of democracy, preferring to keep to a strict scientific argumentation. It is also used loosely, even metaphorically, to indicate a wish to see practitioners interact more closely and less autocratically with rural people. Chambers for example sees the paradigm shift as one from focus on 'things' to a focus on 'people'. If one takes 'paradigm' in its original scientific sense, contrary to Kuhn's assertion, what is presented as a paradigm is evidently not 'incommensurable' with the positivist paradigm.

Chambers and Pretty note that 'scientists must continue their normal science' but 'they will have to learn from and with the farmers' and develop new roles and a 'new professionalism with new concepts, values, methods and behaviour'. This new professionalism is therefore 'not a rejection of modern scientific knowledge a broadening, balancing and up-ending, to give a new primacy to the realities and analyses of poor people themselves..The comfortable certainties of known normal science are then complemented by the exciting unknowns which follow from facilitating analysis by poor rural people and learning from and with them'.

In scientific terms, what is called for is a 'synthesis of opposites' and complementarity: reductionist science will deal with well-defined problems in a situation of low system uncertainty while constructivist science will deal holistically with complex and uncertain problem situations. Methodologically, this has implications: 'a participatory or PRA method can be used as part of an RRA approach, and an RRA method can be used as part of a PRA approach'. Yet the power relations between the two paradigms are not explored.

If the most important feature of normal, positivist, science is its hegemonic worldview, one would expect its paradigm to dominate and impose its logic, that is to instrumentalise participation to ends that are not participatory and to enforce its certainties on the unknowns uncovered by participatory processes. Methodologically, the movement faces two contradictions, both due to a populist stand. There is, on the one hand, the contradiction pointed out by Olivier de Sardan in the 'complex' represented by populist thinking, between the logic of populist ideology and the logic of 'misérabiliste' ideology. While the first emphasises local people's resources, the second emphasises the need to educate them to help them out of their misery.

This contradiction is at the heart of the PRA movement: willingness to respect cognitive and material 'realities' against emancipatory objectives, discourse about local resources and agency but an inability to conceive of local development without resort to outsiders, willingness to use local resources but resort to foreign methodological tools. In terms of 'higher ground' theory, this reflects the contradiction between the assumption of system autonomy and the 'political' wish to steer the system. The recent attempt to go 'beyond farmer-first' is clearly an attempt to move the theory and practice of participation to a post-populist perspective whose terms, as expounded by Scoones and Thompson, are a recognition of the ritual of narratives and the discursive nature of power, the importance of human agency and the refusal of determinism in development. Going beyond farmer-first, in these terms, is going actor-first, but without answering the critique of methodological populism. As will be argued later, it is simply recontextualising contradiction through theoretical elaboration.

The second contradiction is the 'double articulation of discourse' referred to by Laclau in the case of populism, that is the 'dialectical tension between "the people" and classes within the power bloc, and the various ways in which "the people" are articulated with specific classes', cited by. In the case of the PRA movement, as epitomised by Chambers' oft cited categories, this contradiction is expressed in the representation of dichotomic models that obscures these tensions and reduces them to a simplistic opposition between categories, be it social (insiders/outsiders; uppers/lowers; first/last), economic or methodological (conventional research/participatory research; questionnaire survey/PRA) in which sympathy goes instinctively to one side: who could possibly be against participation as a concept? Who could possibly, as a matter of principle, be against the lowers, the insiders or the underclass?. As such, this dichotomic view is reminiscent – not incidentally, as will be argued later – of the old dualist thinking of modernisation theories (modern/traditional, formal/informal processes, developed/ underdeveloped..), which it turns on its head: the attributes of modernity are now seen as 'conventional', even 'traditional' (as 'normal science' is sometimes perhaps ironically referred to) and outgrowing them requires the adoption of a whole new paradigm, but one which is still trapped in the same dualist thinking. If one were to read them in strategic terms, such dichotomies could be considered as pedagogical categories, used to illustrate a point and to achieve political consensus, while assuming that, in fact, there are no definite and deterministic social structures.

Yet these two methodological contradictions need to be replaced in the larger context of post-modern relativism in the social sciences, and more specifically in development, for two reasons. First, the critique of populism, however well argued, fails to address the 'paradigmatic' ambitions of the movement. The critique challenges, so to speak, the 'external validity' of the 'farmer-first' concept around which the PRA movement rallied. Its 'internal validity', if accepted, can probably reject the very notion of populism as a reflection of normal science bias, if it does not integrate it as a legitimate philosophy at all. Second, the movement is very diverse and evolving. The critique of populism – which is, in the case of Olivier de Sardan for example, limited to one Chambers' book – fails to consider the critiques and mitigations coming from within the movement, including attempts to move to a 'beyond farmer-first' agenda. It also fails to address the central issue of the relation between relativism and populism and the more subtle influence it can have on the choice of methods and reasoning. What are therefore the larger, or 'higher ground', contradictions and what are their implications in terms of social theory? An appropriate point of departure is the notion of 'learning', at the heart of the process of change advocated by the movement in its latest theorisation. In itself, the concept of a 'learning community', whether a rural community or a community of 'stakeholders', encapsulates the contradictions between a diversity of influences. The least of them is that 'lowerorder', strongly moral populism is justified by 'higher-order' theories, drawn from the physical sciences, in which moral issues are abstracted out.

REFERENCES

Arce, A., "Beyond State Intervention: Post-modernism and Development" in Frerks, G.E. and den Houden, J.H.B. (eds.) *In Search of the Middle Ground, Essays on the Sociology of Planned Development*, pp.17–40. Wageningen: Wageningen Agricultural University, 1995.

Kloppenburg, J., "Social Theory and the De/Reconstruction of Agricultural Science: Local Knowledge for an Alternative Agriculture", *Rural Sociology*, 1991.

Rosenau, P.M., *Post-Modernism and the Social Sciences: Insights, Inroads and Intrusions*, Princeton, N.J: Princeton University Press, 1992.

9

Virtual Actors and Reality Dysfunctions in Horticultural Research

There is obviously a collective and an institutional dimension to learning. Collectively, 'the learning process leads to change, and debate changes the perceptions of the actors and their readiness to contemplate action. Action is agreed, and those changes which are implemented will therefore represent an accommodation between conflicting views'. At the same time, it is assumed, the process creates a dynamic of 'double loop learning' whereby people become aware of their own learning style, and operate a change in values and procedures that will enable institutions themselves to become 'learning institutions'.

For this, they need to change their feedback mechanisms, abolish hierarchies, decentralise, have fluid and flexible institutional structures, simple decision-making procedures and be more responsive to their environment. But essentially the whole process rests on individuals, on personal change, by learning to adapt one's attitudes and behaviour through self-motivation, resting on the assumed altruism of individuals. 'All learning is ultimately self-learning, which is a very individual process'.

Professional inquiry adopting a learning approach therefore requires autonomous and self-motivated learners. Independently of the institutions in which, or for which, they work, individuals are invited to use their self-critical reflection and their own best judgement. As Chambers points out 'learning is..through doubt, self-examination and willingness to change, seeking self-correcting engagement with dynamic realities and making learning a way of life. In a phrase, this is epistemological awareness, awareness of how we learn, how that affects what we think we know, and how we perceive and distort the realities of others', all because 'in a spirit of pluralism, each person, profession and institution can find an individual pathway for meeting these challenges'.

The view is clearly that institutions are primarily persons, with no special features that transcend individual consciousness or motivation and underlie politics. For exemple the domination to which the lowers/insiders/ underclass are subjected is not considered in terms of economic, social or political 'exploitation' but as cognitive domination. In this sense, much in a post-modernist vein, the personal becomes political and politics is privatised. It becomes not an issue of vested interests but a matter of individual choices and decisions and, historically, the result of socialisation into oppressive educational and professional institutions which, themselves, are detached from private or corporate political interests. Social and, one guesses, economic change comes about through a change in attitudes brought about by appropriate learning processes and action. Structures are no longer fixed, nor really discernible: uppers can become lowers depending on perspective, and the outsider in one system turns into an insider in another.

Concurrently, power is conceptualised in its relational dimension; it is no longer a property of individuals but the outcome of a relation that can both empower and deceive. The powerful are at the mercy of their own self-deception and of the fooling strategies of the powerless who, in turn, can deceive themselves by internalising the powerful's definition. What is depicted is a shadowy world of appearances where nothing is what it seems and everything is language waiting to be decrypted and reification strategies waiting to be revealed.

But while with conscientisation empowerment is assumed to lie with the recovery of 'true' awareness of one's own objective social and economic alienation, a subjectively constructed reality offers only a subjectively constructed power, that is an instantly perishable good, liable to cognitive redefinition by constantly learning actors. Thus while escaping the reification of institutions, the constructivist definition falls into the fetishisation of individual reality.

Most importantly, this conceptualisation of power – and empowerment – ignores two inter-related aspects. First, by taking a discursive view, it focusses on definition and fails to account for the real life impact of power. In short, power is stubborn, and while it may indeed be a social construction, that construction can have real effects – it can still cause exclusion, pain and death, and these cannot be deconstructed without some stretch of the imagination. Second, by focussing on individual realities, empowerment strategies may simply replace the subjugation of overt forms of domination by that of more subtle ones, where self-regulation itself is the means by which power is exercised.

Thus, much in a post-modernist fashion, the analysis of power rests on a self-undermining argument: strategically, if power has no locus or reality in itself it cannot be transferred nor really acquired; methodologically, if power is nowhere it is everywhere and therefore any discursive analysis is bound to eventually legitimise the existing power relations precisely because it recognises none. In a Foucauldian analysis, he points out the 'disciplining discourse' of the state which aims at creating a 'calculating' subject through the enactment of a policy framework that guides and constrains the farmers' 'freedom to choose'. As he says, 'the development of the "calculating" subject is not a function of the direct exercise of power by the state.

Power constitutes this development through the alignment of social norms (economic rationalism), strategies of governance that aim to "free" the individual from the state (deregulation), and a number of programs, techniques and knowledges that infuse every-day practice with the requisite calculative capacity for the development of entrepreneurial and competitive conduct. Through this complex of norms, strategies and techniques, neo-liberalism is able to produce a form of coherence between rationalities of governance and the self-regulating capacities of individuals'. Those farmers who have potential but resist this disciplining are 'adjusted in' through credit and restructuring of their farms. Those who cannot adapt to the policy framework – because of high-risk environment or others – are 'adjusted out'. Thus is struck a new alliance between expertise and politics where, with the 'privatisation of politics', the management of risk is also privatised and participation activists become themselves a tool for the government of individuals through their freedom to choose.

Brown makes a similar point in the case of the PRA movement, which he calls neo-populist. As he says, the results of the 'new professionalism' which PRA practioners call for, may well end up marginalising the interests of the poorest. This is because the public nature of the participatory research techniques, their high opportunity costs to the poor and the bias of institutions, all act to the advantage of the élite.

The shift from PRA to PLA is a further step of the movement towards an integration of broader concerns. It represents an attempt to lay the foundations of a meta-methodology that encompasses all methodologies – past, present and future – espousing (implicitly or explicitly) the 'learning' paradigm. As such, it is simply a re-statement of the cognitivist educational paradigm that, against the behaviourist philosophy of teaching through repetition and reinforcement, sees learning as a proactive process involving the learner as an active goal setter, seeker and creator of knowledge, applying and building information into cognitive structures to solve problems in multiple settings. Though reduced to the role of facilitator, the teacher does not disappear. He simply uses other methods.

Thus Pretty defines the role of the 'facilitating expert' as one of 'helping people in their situation carry out their own study and so achieve something', that is 'changes they regard as improvements'. In this, the learning paradigm reveals its strong pedagogical content. In the PRA movement, this content moves from the educational compound to the social arena, where the cognitive school of educational psychology meets the constructivist school of sociology and the two blend into broader concepts of unpredictable social processes, to produce the notion of a learning community.

The notion is akin to a variety of community development approaches – 'collaborative learning', 'soft system methodology', 'social-environmental learning', 'open space', 'future search', 'holistic management' etc. – and to well established negotiation and facilitation techniques, advocated by different disciplinary fields. Thus with PLA, the appropriative character of PRA, based on its ability to claim tools and concepts from various disciplines and research traditions, moves from the methodological plane to the philosophical plane, but it keeps the same scientific and ideological features. These are looked at in turn.

BUILDING UP STRAW MODELS: SCIENCE OR SCIENTISM

Brown notes that one of the peculiarities of the PRA movement is to confuse 'normal professionalism' and bureaucracy. Similarly, it tends to confuse the functions of science with its dysfunctions. Perhaps at the root of the contradictions of the so-called new paradigm is the simplistic depiction of positivism, wholly associated with normal science, where its shortcomings serve, a contrario, as a foil and a prop to the counter-model. Thus to a naive, often cartoonishly rigid and ahistorical model of positivism, is opposed another naive model of constructivism, in a dichotomic presentation where one picks and chooses selectively those idealised features that best suit the argument, with no nuance or qualification. This dichotomising has two inter-related effects: on the one hand it presents normal science nearly as scientism, that is a naive conception of scientific objectivity and the belief that simple methods, associated with an idealised science, can solve complex problems.

Scientism deals with ideas in the abstract and ignores broader issues of power, economics, politics and all that determines social phenomena and the way we think about them. Ideologues sharing a scientistic vision are all the more dangerous since they can convince themselves that theirs is not an ideological vision, while at the same time serving interests that escape their grasp. They believe in an all-powerful and 'neutral' science, rid of all its ideological elements. As Sokal and Bricmont note, disillusionment with scientistic dogmatism, associated with radical skepticism, can lead to discouragement and rejection of all that was naively believed in before. This, they argue, is what happened after the sixties with a swing to post-modernism. Such rejection leads ultimately to another form of radicalism or, to use the words of Bouveresse, a 'non-knowledge scientism' cited in Policar.

The result of dichotomising, on the other hand, is that the counter-model cannot escape the shortcomings it so categorically dismisses. It replaces one dogmatism by another. The new paradigm assumed to be built in the counter-model is an inverted mirror image of the old one, on which it is patterned and whose elements it moves to a meta-level. But fundamentally we are still in a transfer mode, of software – learning models, forms of social organisation, soft technologies – rather than of hardware – production techniques, hard technologies. We are still told that the 'new' science is ideologically neutral, that it is detached from any political or economic models, that practitioners need not worry about the immediate impact of their work or the wider constraints faced by their rural clients, that things will eventually sort themselves out if we let the 'natural order' run its course.

In this process of change, the arrogance of the appointed 'technical expert' of normal professionalism is replaced by that of the self-proclaimed 'facilitation expert' of a 'normalised' new professionalism. But while the 'normal' professional is still subject to independent assessment of his work, the 'new' professional abides by his own standards. The risk of this 'soft' shift is that the assumed inherent collective features of peasant society which social science took so long to revise may well be creeping back through the window of cognitive psychology and the assumed 'neutral' focus on learning processes.

This transfer of 'software' is concomitant with an increasing communication-information problematic in development thinking which, as will be argued later, itself reflects a failure to break away from, again, a transfer of (higher-order) theories to developing countries. This problematic has perhaps to do with the evolution of post-industrial societies towards information-laden forms of organisation where, lower-order needs having been satisfied, autonomy and participation, together with the use of cybernetic concepts, take on an increased relevance. As Geyer notes, 'when moving from a work-dominated society to an information-dominated one, less centralised planning is a pre-requisite for the very simple reason that the intellectual processes dealing with information are self-steering –and not only self-regulating – and consequently cannot be steered from the outside by definition. In other words: there should be no excessive top-down planning, and science should not get involved in the maintenance of hierarchical power systems'. Yet as argued above, even this selfsteering is not immune to subtle external influences of more diffuse power systems.

MODERNISING PROJECTS

Ideologically, the root of the 'learning' individual in a 'learning' community, lies the old ideal of the creation of a 'new man' in a new type of society, able to rise up to the challenges of the future. This ideal, central to all modernising projects, as Bauer cited in Inkeles and Smith points out, is that of a 'conscious, rational and purposive' man, whose consciousness, rationality and sense of purpose are instrumental to whatever ideology lies behind the modernising project, in a utopian model of society, whether communist or liberal. In a 'learning society', the profile of this new man, as one can figure out from the assumptions of the paradigm, is akin to both the model of the economic entrepreneur – thanks to the agency, knowledge and resourcefulness with which he is endowed – and to the social/communitarian model, thanks to his assumed preparedness to adjust his selfish interests and behaviour to the needs of his community. His cognitive profile is that of a 'cybernetic' individual, able at the same time and concurrently, to gain information from different actors and sources, process it, integrate it into his value system, adjust his system and translate that adjustment into appropriate behaviour, get feedback on that behaviour, process it..and ad infinitum. His mental universe is one of permanent information feedback loops, inducing his value system to produce gradually adapted behaviours. But adapted to what? Information here assumes adjustment of the value system which in turn assumes that of the person's own self-interest to fit the interests of other actors as a result of information.

The possibility of non-automatism, refusal, resistance, in short of nonlinearity, is not considered, even as a working hypothesis. There is thus a contradiction between an assumption of non-linear, 'chaotic' social processes and an assumption of a linear 'predictability' of the cognitive, personal, processes that are assumed to, eventually, produce order or consensus.

Fundamentally, there is a contradiction between, on the one hand, the deeply embedded human desire to seek, by emotional need or purely material design, for stability, regularity, certainty and security of their immediate environment and, on the other hand, the permanent process of adaptation to that environment – itself in a state of permanent disequilibrium in a cybernetic model – assumed as a central tenet of the learning paradigm. The contradiction is basically a political one and relates to the opposition between a vision of politics as personal, peculiar to post-modern thinking, and a more traditional vision of politics as the management of conflictual interests and the necessity of making choices between them, that is ultimately the issue of arbitration or, if one were to use the cybernetics metaphor, who is the cybernetician nudging the learning ghost in the machine towards more timely and accurate feedback? The issue of arbitration, never answered in post-modernism and rarely in the participationist movement, is crucial and needs to be explored, even with the risk of speculating on the intentions, convictions and political inclinations of the theorists.

PHILOSOPHICALLY AND POLITICALLY TOTALISING

The refusal to consider irreconcilable opposition, resistance and contradictions and to recognise the political role of the external agent in the learning process has clear antecedents in political ideologies such as self-management. Lefort, commenting on revolutionary ideologies, pointed out that 'the idea of Revolution as absolute event, foundation of a world in which men would entirely dominate institutions, would agree on all their activities and ends, a world in which power would dissolve itself in the flow of collective decisions, and law in the flow of wills, in which conflict would be eliminated, that idea is secretly linked to the totalitarian representation'.

To be sure, the ideal of a learning community, as suggested by the proponents of the learning paradigm is far removed from the idea of revolution as a guillotine-event, that is the root cause of the communist totalitarian degeneration to which Lefort was referring. Such an idea was used by the state to legitimise its function as the operator of the guillotine and the suppressor of social conflict. By forcefully flattening society into equality, the communist state sought to bring practice at a level with theory and, in so doing, produced a dystopia.

In the proposed paradigm the revolution is embedded in learning itself, and the guillotine of revolution disappears as a decisive moment in history and midwife of radical social change to become the metaphor for a process of change. Yet such vision, much in the vein of post-modernist rationalising away of contradictions through metaphorical images, does not really confront the issue of political power and social conflict. While sharing the totalitarian view of the dissolution of conflict in the collective will, it eschews the issues of social heterogeneity, power inequalities and who is to be the ultimate arbiter in case of irreconcilable oppositions. Except to present learning as a philosopher's stone and dialectical midwife of conflict, it is at the same time its opposite and the means of its resolution. Conflict happens because there was no learning and, therefore, learning does not simply solve conflict, it prevents it from happening, for if it happens this would mean that there was no learning in the first place.

GAP BETWEEN THEORY AND PRACTICE

While the outlines of a learning paradigm give an idea of its theoretical roots, its political content, that is its agenda, is not explicit. What is 'learning' and what is to be learned, by whom and for what purpose? Taken in a general sense, whether explicitly or implicitly, slowly or quickly, willingly or forcibly, formally or informally everybody learn all their life and a learning 'paradigm' seems to make little sense. For practical reasons, learning is quite simply a question of survival and has always been so. In an explicitly educational environment learning has a context and the objectives of specific knowledge and skills acquisition and the teacher is still present.

In a business environment, organisational learning takes place at the request, or with the approval – of business managers, within a clearly defined context of responsibilities, objectives and redistribution. Except in a very general, and metaphorical sense, the rural community is neither a school nor a business enterprise. The assumption of the learning paradigm, one assumes, is that collective learning in the community, as such, is conducive to attitudinal change and therefore to a different, consensual, pattern of action. Meppem and Gill thus define learning as 'the accumulation of insights into system cause and effect by all those with interests in a decision or issue' and in the environmental management and policy domain, as opposed to business management, 'the stakeholder community becomes that entire section of society with interests in the relevant issues'. In a learning environment, they add, 'participants are facilitated away from the need to defend individual perspectives and view points towards a more healthy willingness to listen to and participate in the evolution of more generally shared insights'.

The problem is, of course, that for the purpose of action, 'relevant' systems, issues and decisions need to be defined. While it may be true that ultimately patterns of action are determined by individuals, broader constraints weigh on their judgement and limit their choice. These constraints may be the choices of other individuals in many locations, near and far, and in many domains, whether the local policy maker or the Wall Street stockbroker, some of them may not wish to join in the learning or be 'facilitated away' as one would wish them to because they are happy with the situation as it is.

To assess the potential risks associated with a learning agenda, and banal as this may appear, it is worth considering the difference between theory and its attendant model and practice and its formulation. One obvious difference is that theories do not deal with human beings in the flesh. Theory is, literally, done in a virtual world where human beings are little more than concepts endowed with whatever inherent features suit the theoretical model. The second is that theories refer to models that are conceived out-of-time. Their attendant models are both timeless ideals and the outcomes of an atemporal processes. On the contrary, practice deals with real human beings and their idiosyncrasies, in the real world with all its contradictions.

Politics has to be done in-time, in a frame that is sufficiently short to motivate all human actors, whose time frame is most likely to be their own lifespan, at most. While theory and practice can inform each other, they only do so within a time context to which actors can relate and that gives them hope of achievement, unless, of course, one assumes that they are systematically prepared to relinquish their own objectives for those of their descendants. This time frame needs to be made explicit. To attempt to import a utopian model into the real world is to 'projectise' it, that is to endow it with political intent, fit it into a time frame and confront it to social and cognitive contradictions. What are then the implications of a projectised learning agenda?

Such an agenda, it seems, has not been outlined by the advocates of a learning paradigm. Most probably the answer to the question would be that a learning agenda cannot be projectised, for learning is a life long – and even an inter-generational – process. Similarly, its results themselves can be disregarded. Taking this logic to its limits, its ability to solve problems can give way to its value as a process in itself: learning cannot fail, it can be argued, since its 'failures' serve simply to produce other opportunities for learning. Thus, like a snake biting its own tail, learning is a process that feeds itself. Yet such an answer, again, would eschew two fundamental issues of any agenda. The first is time.

A non projectised agenda, in effect, abstracts out the question of time and, much in a post-modernist fashion, renders obsolete the notion of development as a consciously planned objective. It also undermines one of the central arguments of the learning paradigm: if it is the urgency of the environmental crisis that justifies community learning, then obviously that crisis sets the limits of learning, which needs to be projectised.

The removal of learning from the realm of politics to the meta-level of essential and permanent mental processes thus achieves coherence at the expense of political efficacy. To say that learning is ultimately an individual matter is to suggest it may take time or never occur in the desired direction. After all, any flaws that exist in communities are the result of their own 'learning' style and there is a fundamental contradiction in entrusting the correction of these flaws to learning processes that are their very cause. Besides, whether in a school or in a business model, collective learning is meant to adapt but also to give a competitive advantage over others, if not other individuals then other businesses or schools.

To say that learning can be enhanced is, in effect, to projectise it, that is to say that learning can be taught, and raise the second issue, that is concerning the convenors of the agenda. In the absence of an explicit agenda, one can have nightmares about 'learning' turning into 're-education'and 'learning projects' turning into 're-education camps' led by lesser dictators, in the form of Mr Gadgrind, the Victorian headmaster referred to by Chambers, breathing down the neck of his contingent of active learners, urging them to learn just a little bit faster because aid money is running out and the learning project is to be evaluated soon. To ignore the deeply ingrained human tendency to turn a theory into a method and a method into a recipe is to ignore the risks of degeneration of a utopia into a dystopia.

GAPS AND CONTINUITIES IN DEVELOPMENT THINKING

In concluding, one can start by pointing out, perhaps cynically, two 'original sins' in the search for alternative paradigms; first, the external origin (to the rural world) of all theories of rural promotion and the inability of the theorists to conceive of this promotion without external intervention and, second, the external origin of alternative theories of development. Postmodernism could be said to be untainted by both 'development' thinking and interventionism. Subtly, however, it remains a theory (even in its rejection of any universal theories) that has a conception of development.

Considering these 'sins', the proclaimed break from positivism is itself not a break from a logic of transfer which purports to generalise to developing societies theories that are the outcome of evolving preoccupations of (post) industrial societies. The critique or rejection of the historically constituted western positivist science is itself the historically constituted product of western positivist science. The proclaimed scientific break is inscribed within the same historical continuity. The notion that productivism, served by scientific positivism, is unable to solve the present crisis of western modern societies forces some western thinkers to reconsider the ideology of progress at the root of their modernity and seek new forms of economic and social organisation, mostly to account for the radical change in the form and content of labour.

By rejecting two coherent modernising extremes with totalising objectives, postmodernism occupies a centre that has neither coherence nor objectives or, more precisely, whose objective is to dismantle all certainties, opening the way to alternatives whose objectives and coherence are yet to be defined, solely to be again dismantled. In doing so, it produces another totalising objective, whose rationale is fundamentally 'reactional'. In other words, while one knows what models relativists reject, one is not sure what models they propose. Failing to solve the contradictions produced by this rationale and to make explicit their own political agenda, they open the way to all alternatives, while at the same time depriving themselves of criteria on which to judge them. As Rosenau says, postmodernism leaves 'social science with no bases for knowledge claims and no rationale for choosing between conflicting interpretations'.

POST-MODERN RELATIVISM AND THE SEARCH FOR THE NEW PARADIGM

At this stage of the argument, and before concluding, it is useful to summarise the main shortcomings of post-modernism in the social sciences and its influences on the PRA movement. These relate to its use of the metaphor in the critique of science, the failure to distinguish between issues of scientific method and issues of social power, thus undermining its own argument and, finally, the treatment of politics and economics in terms that make relativism amenable to all sorts of recuperation.

Science and the Metaphor

Perhaps one of the most pervasive post-modernist challenges to science is the use of the metaphor. Ison, Maiteny and Carr for example assert that 'chaos theory, self-organising systems and the complexity sciences bridge the natural, socio-economic and management sciences by serving either as explanations of phenomena or as metaphors which can guide thinking and action'. With other authors, most of the references are made in a loose, suggestive manner, but one which is ambiguous since the arguments that follow build on it and present it as a truth with explanatory power and not a simple metaphor with illustrative intention. The most notable one is the notion of chaos and the sense it conveys that, like the famous effect of a butterfly flapping its wings, minute changes in initial conditions can produce enormous and unpredictable consequences. The implication of this view in philosophical and political terms is further explored in the next section.

As Sokal and Bricmont note, the use of a metaphor in a scientific context is perfectly legitimate if it helps make a point clearer. But it should not be taken as a scientific statement because science and metaphor operate on different registers. Thus Sellier notes the difficulty that science has in dealing with metaphors. 'The scientific debate implies a scientific critique, and the latter is only possible through a discourse that is constructed, the least ambiguous and susceptible to a diversity of poetic interpretations as possible, otherwise one does not know who says what.

The allusive discourse is impermeable to a critique in the framework of the scientific language and, therefore, inaudible as a scientific discourse. Like any discourse, it is of course amenable to literary criticism, whose objectives are entirely different from those of a scientific critique'. As Sokal and Bricmont argue, science is not a text to be read and deconstructed, and exact sciences are not a stock of metaphors ready for instant and unqualified use in social sciences. Concepts such as chaos, uncertainty and non-linearity have a specific meaning that is valid only within a specific theoretical and experimental context. Using them outside that context may lead to absurd conclusions. Sellier, like Sokal and Bricmont, point out the room for backsliding that metaphors afford their authors: when a strict scientific review reveals inconsistencies, they can always fall back on the 'but it's only a metaphor' argument, without in fact solving those problems the metaphor was supposed to help them solve in the first place. Thus, instead of clarifying an issue through analogy, the metaphor adds to the confusion.

Philosophy of Post-modernism

Beyond their metaphorical meaning, the notions of 'chaos', 'uncertainty' and 'non-determinism', when imported into the social sciences, raise a central question: are human beings able to get hold of their future and shape it? Can they write their own history? To use the notion of quantum physics, often suggested by post-modernists, the 'chaotic' view assimilates the unfolding of the future to the dispersion of a wave function holding different potential outcomes, none of which is certain. Like in the analogy of the famous Schrödinger cat in its box, the state of the cat – whether it is dead or alive – cannot be known in advance. It will remain a set of potentialities, only one of which will materialise when the observer opens the box to look inside.

In the social sciences, this vision suggests that the future of societies is beyond their ability to predict or act upon. Like a wave function, it holds 'many worlds', one of which will eventually come into being, as a result of human beings' actions but not their intentions. In development, this view is reflected in the doubts about the validity and efficiency of planning. For this reason, it is argued, social science, especially sociology, should function on the principles of meteorology and abandon its predictive ambition. What is suggested is therefore an ontologic powerlessness of men to determine the future they want and, as many critics of post-modern philosophy have argued, a deeply pessimistic view under an apparently libertarian message.

The 'chaotic' view has much deeper philosophical and methodological implications than its proponents are willing to admit and, as far as PRA theorists who refer to it are concerned, its commensurability with normal science is seriously in doubt. What is in question is, fundamentally, the meaning of 'science'. Philosophically, what it suggests, in fact, is to integrate the unknowable into the realm of knowledge and make it a guiding principle of scientific reasoning. Methodologically, this means that, as Olivier de Sardan wrote in another context, 'the unexplained is explained by the inexplicable'. This is a most radical challenge to the classical schools of sociology which consider social systems as historically determined.

Adam Smith and Karl Marx saw respectively an unknown but knowable 'invisible hand' or 'alien force' as the resultant of human intercourse, and power as resting with nature and not with God. By discovering the laws of nature and those of human conduct, science could empower men to control their destiny. The challenge to these views is not new. The recent history of social science, concurrently with the decline of the narrow positivistic view, is precisely a move away from the ambition to predict social phenomena. Similarly, while not adopting uncertainty as a philosophical principle, science has recognised it and dealt with it: what is 'risk management' in economics if not dealing with uncertainty?

But the post-modern challenge, if one accepts a paradigm espousing the 'chaotic' view, is of another nature altogether: the rule of 'chaos' is, so to speak, a reinstatement of power into the hands of God, and the process through which His will is approached, but in essence impossible to know, is through personal introspection, learning and participation, are thus turned into the rituals of a new religion. Where Kuhn saw a paradigm as a consensus among the scientific community as to the way its members view the world, the new paradigm enlarges that community to a much wider audience. Where Kuhn saw the validity of a new paradigm in its ability to solve problems better than the old one, the proposed paradigm makes problems redundant by 'deconstructing' them as mere fabrications of science. Problems, we are told, are endemic and their solution increases ignorance instead of knowledge; thus 'problem-solving' in the new paradigm becomes a veritable quest for the Holy Grail, always pursued but never reached.

Where science posits the hope of an attainable truth, the new paradigm suggests that the hope resides in the attempt, not in the achievement of the goal. To men who would seize control of their destiny, post-modernism tells them this is a laudable but vain hope. This quasi-religious dimension is present in the notion of time and/or projectising discussed earlier in the case of learning. The refusal to projectise reminds one of the religious belief in the ultimate achievement of some mythical objective and the role that practitioners acting as missionaries play in sowing the seeds of change, even if the horizon of that change is far removed. This dimension is also transparent in the refusal to consider arbitration and external, godly, intervention: God, it is assumed, cannot meddle in our affairs or bypass our agency by protecting us from the consequences of our actions.

Radical as it may sound, this view ignores both the history of societies and human nature. Basically there is a confusion between the field of knowledge and the field of action in planning, more specifically the extent to which planning is likely to achieve man's will without prior perfect knowledge.

Politically, although no implication of this position is drawn by the critics of planning, their analysis conveys an unarticulated sense of laissez faire. But here again, the philosophical dismissal of planning, as a means for men to shape their future, is a denial of history. While development planning may have dismally failed in many instances in the Third World, the spectacular development of the post-war period in Europe (France, Germany) and in Asia (Japan, Korea) is proof enough that determined action, backed by economic and human resources and a vision of the future, can lead men where they want to go. Whether they like the results is a matter of dynamics of human nature, not a fundamental inability to turn intentions into reality.

Methodology of Post-modernism

The rationale of post-modernism is to systematically turn them into definite truths, even in its rejection of any universal truth, and aggregate them to make claims to a philosophy. The meta-theories it produces – its own meta-discourse of knowledge about knowledge, thinking about thinking, learning to learn, communicating about communication, the politics of politic etc. reproduces the contradictions instead of solving them. There is therefore, as Hammersley and Gomm note, a built-in 'self-undermining internal inconsistency of relativism' which leads relativists to 'oscillate between undiscriminating tolerance and ideological dogmatism', one time seeming to accept every interpretation as valid in its own right, the other retreating behind the incommensurability of paradigms. Thus relativists say that we live in a world of multiple realities but, of course, 'the argument that this is the nature of the world itself constitutes a claim to universal validity'.

As has long been recognised, 'the desire to be value-free is in fact a value in itself. Hence to attempt to demonstrate that any science is value-free is a pointless exercise'. Post-modernists tell us that development is a form of discourse, but to say this is also to produce a discourse about development which, in a way, is intellectually, if not morally, even more untenable since it continues to feed at the same trough of development studies, if not of development aid. They tell us that knowledge is a form of discourse and there is no 'true knowledge' that could be enjoyed by somebody sitting in a transcendental watch tower, but this is a statement made from a transcendental watch tower.

Science is criticised for its methodological reductionism but post-modernism itself resorts to conceptual reductionism in that it reduces social and cognitive processes to a single universal law of relativism, true in all bodies of science. The advocates of the new paradigm tell us that the future of communities lies in their ability to learn how to learn, but this is an ambiguous statement since it suggests that they are in a position to judge what is true learning, if not to teach it.

Politics of Post-modernism

An important question to ask about the implications of post-modernist thought is that of the political and economic models it conveys. Its political agenda is, at the very least, ambiguous and this ambiguity is also reflected in actor-oriented sociology. Some see in it an attempt to lay the scientific and philosophical foundations of liberalism, by asserting the pre-eminence of the individual over broader social structures, reduced to simple 'constructions'. Human agency, it is thought, can be interpreted as the sociological foundation of entrepreneurship and the 'cognitive respect' due to individuals is also an invitation to respect their individual freedom to choose, over any other consideration.

Although there is not one single model of politics that can be deducted from post-modern theory, its philosophy points to one central notion on which politics should be founded. Against the Marxist overdetermination of individual subjectivity by the economy, post-modernists argue that subjectivity is fragmented by the individual's multiple positions, defined by his/her different social relations and the discourses which they produce. Thus, so to speak, the individual's is a 'variable geometry' of subjectivity, which can never be fixed. The implication of this is that politics – as a form of organised subjectivity – can no longer be thought of in terms of the traditional, and universal, notions of left and right, socialism or liberalism, and cannot crystallise around parties or concepts as reductionist as a class. Politics descends to the social sphere where it is articulated by new social movements whose identity is rooted in various subjectivities.

Fundamentally, the politics of postmodernism is based on the twin claims to the rights to difference and equality. Such vision leads postmodernists to see the marginal, the informal and the oppositional as powerful sources of creativity: thus informal markets, ethnic minorities, ecological groups, squatter groups, homosexuals, etc. are seen as the main carriers of the potential for social change. Their agency is basically one of a power of resistance to any homogenising attempts of the state. These movements are not interested in access to state power but in the creation of an autonomous space. They are self-producing and self-organising; their concerns are embedded in subaltern knowledges and expressed in cultural terms. As Aronowitz cited in Slater says, 'new social movements speak in post-modern voices; they enter the national and international political arena speaking a language of localism and regionalism'.

In development, it is a contradiction that it is precisely when obstacles to overt political action have lessened and opportunities for organisation have opened in most developing countries that development theorists choose to abandon the fields of politics and political economy in favour of an ambiguous discourse which tends to dissolve Third World specificities by absorbing and neutralising the realities of domination. Differences, whether local, sexual or communitarian, are subtly aggregated in a framework that neutralises them: the potentially destabilising 'other' is integrated in the form of an undifferentiated and interchangeable whole. The post-modern discourse is thus ambivalent: the deconstruction of established concepts is a potentially political and empowering element, but it is undermined at the root by the denial of power relations, or more exactly by the dissolution of power relations into language games or their trivialisation by reference to extreme, pathological examples.

Perhaps at the root of post-modernism's incapacity to articulate a coherent political project is the logic of trying to raise common sense to the status of a theory, if not a philosophy. The political left and right were founded on 'motivational' hypotheses about human nature: the assumed orientation towards community, social justice, solidarity or, on the other hand, individual gains and rights. The attempts to go beyond this divide are founded on concepts that are more a statement of personal attributes than an indication of human aspiration.

Notions like 'agency' and 'learning' are intuitively felt to be true. They are simply part of being endowed with consciousness, motivation and survival instinct, that is ultimately part of being human. Trying to raise them to the status of scientific concepts and build a theory around them eventually fails to explain why these notions were not able to build a better world than the one we live in. As analytical concepts in the field of knowledge, they give precious insights into the 'micro-physics' of power; in the field of action, they fail to produce a strategy for the correction of those phenomena they highlight.

A strategy that purports to build on those very notions that produced injustice cannot produce justice, if at the same time it dismisses determined action in favour of more justice as mere interventionism or as a vain modernisation attempt. What it does is redefine semantically the attributes of the powerless and romanticise the condition of the marginalised groups while ignoring the subaltern position in which they live. It turns them into capable actors, while at the same time ignoring the fundamental roots of their powerlessness and suggesting that their agency should be left to its own devices. In short, theorising common sense produces a politics of the status quo. As Barnhart says, 'ultimately, while post-modernism attempts to deconstruct macrotheoretical models, it fails to account for phenomena of social change the models were constructed to analyse'.

By trying to lay the foundation of a new form of politics, and a new model of democracy, postmodernism lays itself open to criticism from both left and right. The central critique of 'order' and the emphasis on non-determinism can of course refer to anarchism. But there is a danger that it is used politically to justify any orders, including a 'new order'.

The right sees in its 'deconstructionist' style an attempt to undermine traditional western values. The critique from the radical political left is severe: abandonment of critical theory, extreme relativism, denial of the 'reality' of social and economic domination, including relations of exploitation etc. One of the most important criticisms is that it promotes notions such as the specificity of women's work and of cultural groups, communautarism and the rejection of universalism, central to the agendas of most religious fundamentalist groups and the political extreme right. To MacDonald, post-modernism boils down to 'middle class liberalism with all its attendant features of naivety, irrationalism and idealism' which 'despite its left veneer..ultimately leads to reactionary conclusions'.

Nanda argues that post-modernist philosophy, with its emphasis on cultural relativism can easily be used to condone the excesses of the Hindu fundamentalists in India. She makes a damning statement on the temptation of the political left to verse into relativism and associates the move to higher theories with a decline of its political efficacy. 'What do left intellectuals do when they know that they are too marginalised to change the world? They get busy interpreting the world, of course. And interpreting how we interpret the world, and how the non-Western "others" interpret it, and how we interpret others' interpretations..and ad infinitum. The interpretive turn allows the left to create in discourse what it is unable to realise in the rough and tumble of real politics: a world where all ideologies have been deconstructed, revealed, and readied for overthrow; a world where all can live by their own lights'. In a more moderate view, Booth notes that, rid of its intellectualising jargon, it is but a reflection of 'a relatively widespread mood', born out of an increasing awareness of the limitations of over-deterministic development theories. There is little doubt, however, that as a strategy, or rather a set of strategies, post-modern relativism offers little direction, except to emphasise that this directionlessness is liberating and a source of challenge.

Taken as a critique of modernisation theories, post-modernist relativism seems little more than a rehashing and systematising of old arguments. Post-modernists have often been branded 'armchair radicals' in that their critiques focus on changing ways of thinking rather than calling for action based on these changes. Such critique is echoed in the field of participatory research by the often heard depiction of PRA promoters as 'yuppy anthropologists', attempting a 'quick and dirty' synthesis of various old methods of qualitative research. Post-modernism, PRA, and more broadly the participationist movement, are also a reflection of a 'widespread mood' in the development field. The proposed participatory paradigm, sharing post-modern relativism and having failed to provide a sense of direction about practice, ultimately suggests that participation is a question of 'state of mind' rather than of any particular method. Ultimately, the issue is one of personal convictions and commitment. Whether these amount to a new scientific paradigm remains to be seen.

Science of Post-modernism

Whichever way one looks at it, the post-modern critique of science – and to a large extent that of the advocates of a new paradigm – is so overwhelming as to be a veritable indictment. Despite passing mitigations, its general thrust is that science, portent of the project of the Enlightenment, is omnipresent, omnipotent and overbearing. Politics is assumed to be guided by science, either directly – because politics is advised by scientists – or indirectly – because political decisions are built following a scientific rationale. What science does not do directly, it inspires. Professional politicians themselves may be seen as, first of all, science-trained individuals.

Thus science is presented as a 'total culprit', responsible for what happens as well as for what does not happen, because it could have made it happen. Although critics do not go so far as to put scientists on trial, this view, if accepted, raises the central isssue of the collective social responsibility and accountability of normal scientists. Such a view embodies perhaps the most basic confusion in the post-modern critique of science, that between the critique of the institution, i.e., the field of sociology of science, and the critique of the scientific method and knowledge, i.e., the field of epistemology.

While the body of scientific knowledge is indeed also the product of the institution, post-modernists are too swift in 'throwing out the baby with the bath water'. They do so fundamentally by refusing to make a distinction between science and politics, between science as a method and the scientist as a person and a citizen, between 'reality' and the validity of its social representation. Knowledge here is viewed as, basically, a form of discourse creating its own subjective world. Such a radical view conveys the idea that all knowledge is equally valid, or equally suspect. As Sagan says, 'the well-meaning contention that all ideas have equal merit seems to me little different from the disastrous contention that no ideas have any merit'. PRA theorists do not fall into this reasoning, but they entertain the ambiguity by developing a discourse which enshrines local knowledge while at the same time relativising the value of scientific knowledge. There is a contradiction in considering critically, and dismissing, scientific knowledge as being socially constructed while, at the same time, never questioning the validity of local knowledge.

As Molnar et al. say in their response to Kloppenburg, 'there is only one science' and that is 'a way of accumulating knowledge in intersubjectively testable ways'. While some other ways of knowing exist, and may be superior to science in some contexts, they are not science. The issue is, fundamentally, one of making a distinction between knowledge founded on interpretation and knowledge constructed on systematic observation and experimentation. Local knowledge has both forms. It is neither superior nor inferior to scientific knowledge; it is adapted to specific situations which science has not been able to deal with. It can also, clearly, become maladapted as local situations change and local interpretations are no longer valid, if they ever were.

Economic and environmental pressures are brought to bear on them and stretch them to the limit of their relevance. In this case, much as it may sound politically correct, not questioning the validity of local knowledge may well become the surest way of burying local communities under their present problems. On the issue of science and politics, Aronowitz, for example, justifies the relevance of scientific studies by stating that 'what is at issue..is the question of scientific citizenship. The wide influence of science and its applications make it clear that science policy is too important to be left exclusively to scientists, their corporate patrons and politicians'. This issue raises, indirectly, the question of whether science should function according to the rule of democracy or whether it should submit itself to democratic control. In other words, should democracy be exercised in science or on science? Through the exercise of democracy, control is exercised and dominant social values are integrated into science. But integrated how? How is this scientific citizenship translated in practice of science? Should this practice reflect social values or should simply scientific research policy be decided democratically?

Latour notes that there are differences between research, that is the institutional domain of science, and science itself. While the former is open to influences from its socio-political context and to the foibles of researchers, the latter is a body of knowledge and methods that can serve politics but whose relation to politics is difficult to establish. Molnar et al. also emphasize that 'science is a human enterprise' and the institution of science has its share of shameful episodes. As all human enterprises, science has shortcomings that reflect the limits of society. Scientific institutions need to be changed to make them more responsive to economic, social and cultural changes. When their products are the source of inequalities, society must intervene, but this is an issue of politics, not of science.

Scientists can express their political preferences but these are theirs, not science's. As human beings, and citizens, scientists are sometimes lured by personal gain and glory. Sokal and Bricmont note that 'scientists are only human beings and they are immune to neither fashion nor adulation'. They may ignore the scientific and ethical rules of their practice and err on the side of career concerns and ambitions. But one of the legacies of the Enlightenment, they remind us, is also skepticism and a distrust towards sacred texts and arguments of authority, and scientists can exercise them when it comes to science.

Social constructivism has a role to play in operationalising scientific skepticism. For example, the process of consensus building about what is 'valid' and 'objective' research among the scientific community can only work well if there is a wide and substantive dialogue in that community. In this sense, social constructivist analysis can help redress biases and be conducive to a stronger form of objectivity. There is however a difference between specific skepticism and radical skepticism, and the use of social constructivism ceases when it lapses into the latter. The first is interesting and can be refuted, the second is irrefutable but is less interesting because, precisely, it aims at universalism.

Local populations must have a say in scientific research, but this is essentially an issue of democratic control, not of values. This control is best advanced by the promotion of their democratic rights, that is by replacing the debate firmly in the field of politics, in which scientists, as citizens have, of course, a say. But much as it may sound generous and altruistic, scientists may fail farmers by relinquishing the skills for which they have been trained.

REFERENCES

Checkland, P. and Scholes, J., *Soft System Methodology in Action*, New York: John Wiley, 1990.

Giddens, A., "Central Problems in Social Theory: Action, Structure and Contradiction", *Social Analysis*, Berkley: University of California Press, 1984.

Gorz, A., *Farewell to the Working Class: An Essay on Post-Industrial Socialism*, London: Pluto Press, 1980.

10

Assessment of Horticultural Research

Agricultural research generates many types of outputs. These include technologies embodied in a physical object (e.g., improved seeds), management tools and practices, information, and improved human resources. These outputs affect the environment of research institutes (through training and partnership building) and research clientele (through technologies and information generated), which ultimately impact the indicators of research goals. In the case of the CGIAR, the goals are to alleviaie poverty and protect natural resources so as to achieve sustainable food security. Thus, we ideally and ultimately would want to assess impacts of CGIAR programs in terms of how they contribute to meeting those goals. However, these types of fundamental goals related to human welfare are several steps away from intermediate goals and from objectives of the specific research projects/programs. Thus, intermediate goals, such as increased capacity of agricultural research organisations and increased sustainable agricultural productivity through the development of or promotion of improved technologies are more easily measured in terms of cause and effect, and impacts.

Usually, it is difficult to evaluate impacts in terms of the ultimate broader goals of poverty alleviation and environmental sustainability. Thus, research evaluation tends to concentrate on impacts in terms of measurable intermediate goals or objectives. In the above case, it would attempt to evaluate whether the research in question resulted in technologies, management strategies and capacity strengthening that lead to more agricultural production per hectare at lower cost per unit of output and in a more sustainable fashion for more people. By assumption one would expect that increased and more sustainable food production by farmers would lead to an increase in their welfare and possibly, through price reductions for basic food, to an increase in consumer welfare.

Impact assessments used in accountability exercises to justify a past program activities should evaluate impacts in terms of all the goals and objectives that were set when the project, activity or program was established. Thus, if the research goal included ensuring resource conservation in addition to productivity enhancement, then impact assessment should measure the positive or negative interactions between productivity increments and resource degradation that may have resulted from new technology to derive an overall evaluation of research impacts. In cases where IA is used to provide understanding for future planning, it is not as critical that it responds to past established goals and objectives. Similarly, where it is used in terms of understanding impacts on local research and development capacity of organisations, a different framework may be called for in the activity.

Impact assessment is an established practice in public-goods investment projects and programs such as infrastructure, health, education, transportation, and urban development projects. The International Association for Impact Assessment (IAIA), among other things, organises an annual conference that this year was attended by nearly 600 people from 80 countries. There is a rich literature covering this wide range of development projects and distinguishes "impact assessment" on disciplinary lines as:

— Environmental impact assessment (EIA)
— Social impact assessment (SIA)
— Environmental health impact assessment (EHIA)
— Risk assessment
— Strategic environmental assessment (SEA)
— Economic impact evaluation (EIE)

Recently, emphasis has also been given to Institutional Impact Assessment. Many countries have guidelines for one or more of these types of IA. The IIED's "Directory of Impact Assessment Guidelines" contains over 800 bibliographic references and abstracts for more than 90 countries and 45 international development agencies. Very few of these deal explicitly with agriculture and none identified deal with agricultural research. Thus, compared with other development projects, agricultural research has still not entered the mainstream of "impact assessment" literature. Even after more than four decades since the publication of first studies of the economic impacts of agricultural research, and hundreds of impact studies thereafter, impact assessment remains an uncommon exercise in agricultural research programs. Many researchers and managers of agricultural research are not convinced of the usefulness of impact evaluation.

There are many reasons for the reluctance on the part of research managers to undertake impact assessment of agricultural research. One is that research resources are scarce and most research managers are of the opinion that the benefits are obvious anyway. Given the high opportunity cost of doing IA, research managers may feel that they should simply get on with the research. Perhaps another reason may be the fear that evaluation of research results would produce unfavorable benefit-cost ratios. Also, few impact assessments provide clear guidelines for future action, making them less valuable as research tools. Methodological difficulties in estimating the benefits from some types of research, may yet be another reason for the reluctance to conduct impact assessment.

Several attempts are underway at national and international levels to change this perspective and make impact assessment a routine practice in agricultural research institutes. A general overview of different approaches to research evaluation and IA of agricultural research and of activities of organisations involved is provided by Alex. This workshop is one example of such attempts to increase the knowledge and awareness among research managers and decision-makers on available tools, methods and practices of impact assessment. The discussion in the following sections hopefully will help towards increasing this awareness.

Impact assessment is done for several practical reasons: (1) Accountability - to evaluate how well we have done in the past, to report to stakeholders on the return to their investment, and to underpin political support for continued investment; (2) Improving program design and implementation - to learn lessons from past that can be applied in improving efficiency of research programs; and (3) Planning and prioritising - to assess likely future impacts of institutional actions and investment of resources, with results being used in resource allocation and prioritising future programs and activities, and designing policies, programs and projects.

Impact assessment is viewed as a type of research evaluation. Research evaluation involves judging, appraising, or determining the worth, value or quality of research (proposed, ongoing or completed) and is done in terms of its relevance, effectiveness, efficiency and impact. Impact assessment can be carried out at different levels of aggregation—individual research projects, specific research programs, or the research system as a whole—depending on the objectives and type of the exercise. There are several stages of research evaluation along a time continuum, and impact assessment can be viewed as occurring in the design and post-adoption stages at different levels of research system as depicted in Figure 1.

Looking at the past, we have ex post impact assessment and evaluation of performance, achievements and impacts. This is most suitable at the aggregate system level or a research program level. The resulting information is used in accounting for the past use of resources; and as a useful input for research policy making and planning for the future. During the present stage along the continuum, there is monitoring and evaluation of on-going research activities aimed at providing information to guide present activities and revision of ongoing plans. This is usually done at a research project level. Looking towards the future, there is ex ante impact assessment of likely future environments and of expected impacts from research. Ex ante impact assessment is applied to assist in decisions on approval and funding of research and is generally done at a project level. However, at a research system level, ex ante impact assessment is also conducted to rank research programs and set priorities for resource allocation purposes.

While we have a time continuum from past research outputs, through present research activities and on to expected future activities and output, the evaluation and assessment activities and their outputs fit into a broader dynamic evaluation cycle, in which the results of monitoring and evaluation (M&E) at a project level continuously feed back into the impact assessment and priority setting at the program- and system-levels. Thus, all three points along the time continuum involve evaluation or assessment to provide information for planning future priorities, strategies and activities. Even in the case of ex post impact evaluations for accountability purposes, the resulting outputs can be used to adjust future funding allocations and personnel assignments.

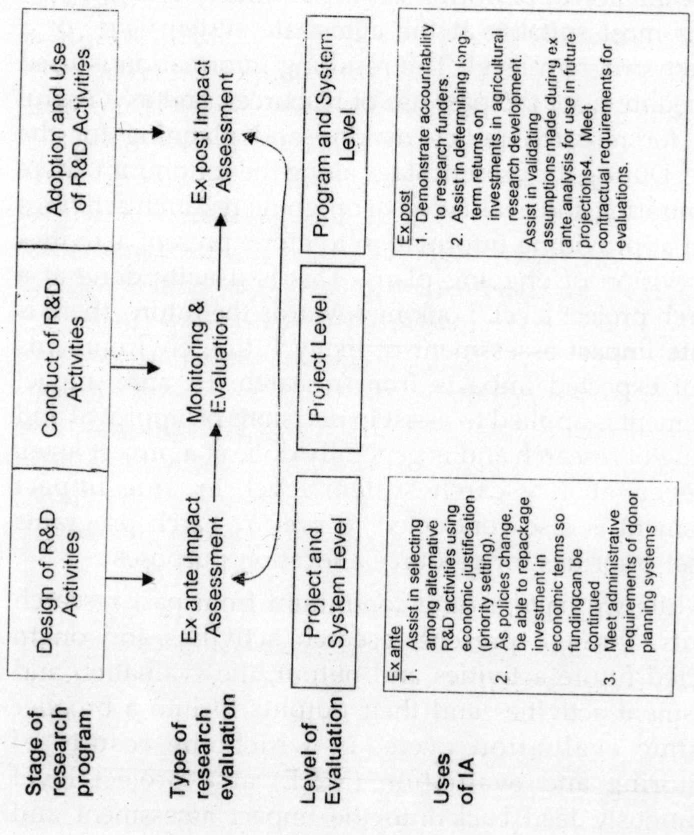

Figure 1: Types and uses of research evaluation by stages of research program implementation

Impact assessments should be done when needed and as required. Thus, we may do an IA when we are planning future research (ex ante), when we need to show what past research has yielded in terms of social and economic benefits (ex post), when we are applying for new or continued funding (both ex ante and ex post).

Most commonly, IAs are done because decision makers and research funders require them as a pre-condition for research support. The significant expansion of impact studies, both ex ante and ex post, in the past decade or so have been motivated by this factor.

This requirement has the added benefit that it creates institutional capacity to carry out impact assessment in developing countries. To be useful as a tool for research planning and policy making, this initial capacity should be further developed and used for "self-evaluation" purposes. The second generation of impact studies that we see in recent years, especially in industrialised countries, reflect this trend and the growing importance placed on the objective of research efficiency as against accountability.

Increasingly, impact studies are undertaken with the aim of:

(1) helping managers provide better and more convincing advice on strategic decisions about future R&D investment;

(2) making scientists and researchers aware of the broader implications (or lack thereof) of their research;

(3) identifying weak links in the research-to-impact pathways; and

(4) better informing research managers on the complementarities and trade-offs between different activities within a research programme.

Impact assessment should be used as a tool for research planning and institution building. To be useful as a learning tool, (ex post) IA should be undertaken at least once in every five years for each program.

AGRICULTURAL RESEARCH IMPACT ASSESSMENT

Agricultural research systems worldwide undergo various forms of reviews and evaluations as is appropriate to their perceived importance as public sector investments. Some of these evaluations are administrative or financial, others are formal but subjective "peer" reviews and ratings. Some reviews are formal impact evaluations that give quantitative measures of impact assessment, which may be more acceptable to funding agencies and/or useful in fairly sophisticated agricultural research planning. We assume that these are the concerns of this workshop. The following sources of examples and information provide insights on the state of the art practice related to economic, social, environmental and institutional impact assessment.

Economic Impact Assessment

When it comes to assessment of the impacts of agricultural research, the main quantitative work has been done in economic impact assessment. These impact studies range in scope and depth of evaluation from partial impact studies to comprehensive assessment of economic impacts. Partial impact assessment studies are more concerned with what the direct products have actually led to in farmers' fields (or where ever else research is applied). One popular type of partial impact assessment is adoption studies (or pre-impact studies) that look at the effects of new technology such as the spread of modern plant varieties on farm productivity and farmers' welfare. Economic impact assessments of the more comprehensive types look beyond mere yield and crop intensities to the wider economic effects of the adoption of new technology. These studies generally estimate the economic benefits produced by research in relation to associated costs and estimate a rate of return to research investments. Unlike adoption studies, which are done only ex post, economic studies are undertaken both ex ante and ex post.

The state of the art in adoption and economic studies, it is important to point out that most of these types of IA studies do not look at the full general equilibrium (GE) effects of research including the indirect impacts on product and factor markets. These effects are realised not only within a country but also inter-regionally if interregional labor markets exist. A few studies have tried to estimate these GE effects. However, there are no GE models developed that are both practical and detailed enough to provide much guidance for evaluating research impacts. Isolating and attributing the impact value specifically to the research component remains a difficult issue to tackle for agricultural research IA. The approaches discussed below assess research impacts in a partial equilibrium setting.

State of the art in adoption studies

Adoption studies generally trace the results of plant and animal innovations from the research station or on-farm trials through networks of adopters. This type of evaluation analyses the underlying patterns of adoption and the use of new practices using statistical and econometric tools. Adoption surveys are frequently used to see if farmers (or other research clients) are using (or not using) improved technology and to look at its effects on farm production, to measure client satisfaction (or dissatisfaction) of research results, and to determine how research activities can be reoriented to make technologies more useful. These studies attempt to determine why a technology is or is not being used and sometimes to compare the benefits of old versus new technologies. An excellent extract of the relevant literature on the theoretical concepts and factors underlying adoption of improved technology and the process of innovation has been put together by Kuby.

Adoption studies are more common in developing countries than the comprehensive economic studies. The inventory of impact assessment studies in the ASARECA region compiled by Omiti, Waithaka and Mwangi indicates an impressive list of adoption and diffusion studies undertaken in this region. There is also an impressive body of literature on adoption studies in other parts of developing world as well. Feder et al., and Feder and Umali surveyed much of the literature in this area. CIMMYT's publication on this topic provides a good resource guide that can be used for designing surveys to conduct adoption studies.

Impact assessments in the CGIAR System also began with adoption and diffusion studies on semi-dwarf rice and wheat developed by IRRI and CIMMYT. Using national level statistics, the USAID documented the development and global spread of HYVs of rice and wheat. Such publications began in 1969 with data from the 1965-66 crop year and continued until 1986. These USAID supported studies were later extended to hybrid maize. CIMMYT had undertaken a series of adoption studies in the 1970s. In the early 1990s, CIMMYT undertook much broader and in-depth studies towards updating the adoption data for maize and wheat. An updated overview of the impact of rice research is provided by Pingali and Hossain. Recently, under the germplasm impact assessment work of IAEG, such adoption studies are being undertaken at all crop-based CGIAR centers in close collaboration with their national program partners using a common conceptual framework and methodology.

The adoption and diffusion studies form the first step towards building an institutional capacity to undertake comprehensive impact assessments (and therefore sometimes referred as pre-impact studies). The data and results of adoption studies provide the baseline data for the evaluation of technology impacts on productivity, income, environment, equity and other goals. Analysis of constraints to technology adoption are also important in increasing research efficiency, as well as in designing policy, programs and institutional reforms to enhance technology adoption. It is therefore important from an institutional perspective to conduct regular adoption studies and surveys in defined areas to build a credible data set for comprehensive impact studies. While economists have initiated, and may still dominate the agricultural technology adoption studies, anthropologists and sociologists have enriched the adoption analysis through more careful observations and more attention to less quantifiable constraining and facilitating factors.

State of the art in economic studies

Economic studies include studies that estimate economic benefits and measure economic rates of return (and to some extent the distribution of these returns) associated with innovations coming out of research investment. Since the pioneering efforts of Schultz and Griliches, and despite some of the measurement problems discussed in the literature (e.g. separating research impacts from extension and private sector investments and attribution problem mentioned above), there have been significant developments in the methodology of evaluating the economic impact of agricultural research. Today there is a continuum of methodologies, ranging from "back of the envelope" accounting of crude costs and benefits to more sophisticated costbenefit, economic-surplus, and econometric analyses. The literature on economic impact studies also includes a wide range of levels of impact analysis, from aggregate, national level to program and project level. The econometric approach of estimating research productivity and the total factor productivity

(TFP) analysis are best suited at the very aggregate-level of impact assessment. On the other hand, the economic surplus and cost-benefit studies are most suitable at the level of individual research program. The cost-benefit and modified economic surplus methods are also common in ex ante economic assessments.

An economic study at the research program level helps:

— Identify what the technical and economic outcomes of the R&D were (will be) and help identify why such outcomes were (will be) achieved; and
— Assess the welfare effect of the R&D by comparing it with the level of welfare that would have arisen in the absence of the research project.

Figure 2 summarises the task of an economic study. Ideally an economic evaluation of R&D project or program is a process and way of thinking about the economic importance of R&D. It should be all inclusive, assessing all the benefits and costs of expenditure on R&D. These include:

— Direct financial impacts through changes in input costs (quantity and/or prices), output price (including quality effects) and quantity produced;
— The costs and revenue of products or activities replaced by the new technology;
— Indirect financial and economic impacts through other spillovers;
— Growth linkage effects; and
— Externalities through benefits and costs imposed on others. Of particular importance for agricultural research projects are the environmental benefits and costs, such as land and water conservation or degradation.

Much economic research has focused on measuring the direct economic costs and benefits listed in categories 1 and

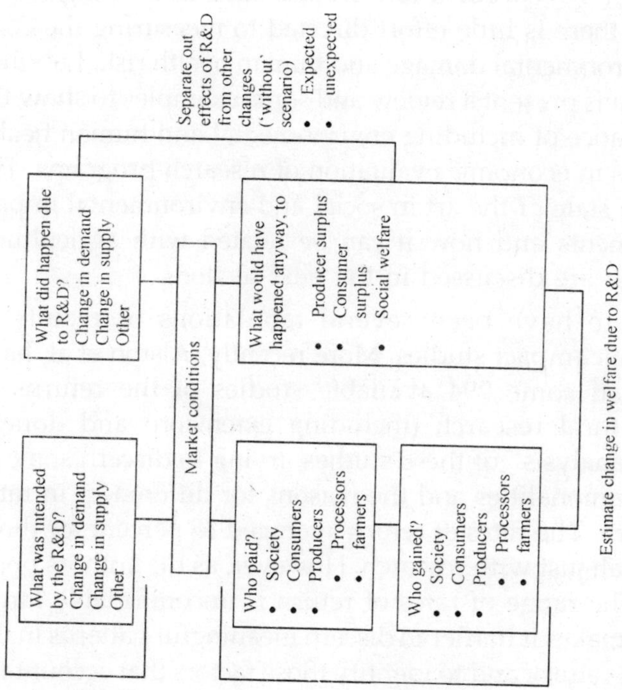

Figure 2: The Task of Economic Impact Assessment of Research Investment

2, and to some extent those in category 3. There are very few studies that incorporate social and environmental externality costs in to the calculation of economic surplus. The ISNAR manual by Janssen and Kissi is one of the few examples of incorporating environmental costs and benefits in an ex ante impact assessment of agricultural research. Except for a few studies such as by Pingali and Roger, there is little effort directed to measuring the costs of environmental damage and human health risk. Lubulwa and Davis present a review and some examples to show the importance of including environmental and human health impacts in economic evaluation of research programs. The current state of the art in social and environmental impact assessments and how it can be related with agricultural research are discussed in the next sections.

There have been several tabulations of results of economic impact studies. More recently, Alston et al. have reviewed some 294 available studies of the returns to agricultural research (including extension) and done a "meta analysis" of these studies, trying to discern some of the commonalities and the reasons for differences in rates of return. The rates of return averaged 88 percent for those that dealt just with research. However, as the authors point out, "The range of rates of return is uncomfortably large, which makes it harder to discern meaningful patterns in the rates of return, and to identify those factors that account for the systematic variation in the evidence".

The authors found evidence that longer production cycles were associated with lower rates of return and that significantly lower rates of return were associated with natural resources management compared with other categories. An interesting, but expected result of the study was the evidence that the characteristics of the research evaluation itself, and the analyst conducting the evaluation, were found to have important systematic effects on the estimated rates of return.

The state of the art in economic impact assessment has recently moved forward considerably by the publication of Science Under Scarcity, a book prepared by Alston, Norton and Pardey, as part of an ISNAR initiative (and recently republished by CABI). If the methods elaborated in this book find their way into routine practice, the state of the art of research evaluation would be advanced profoundly.

Methods developed, however, relate particularly to commodity research, and mostly to crop varieties. Traxler and Byerlee and Joshi and Bantilan are some of the few examples of an impact evaluation of crop management research. Assessment methods and analyses of livestock, fisheries and agro-forestry research impacts have received less attention in the past, although this trend is changing rapidly, especially with impact assessment studies being conducted by CGIAR centers such as ICLARM, ILRI, ICRAF and CIFOR in collaboration with national research programs. The methodological challenges to assess the impact of social science and policy research, institution building and natural resources management research remain very large despite recent efforts by may research institutes to initiate IA of these types of research. The sourcebook by Horton et al. provide a list of resources and reference materials that can aid researchers in the economic evaluation and impact assessment process.

Considering the evolution of theory and methods to evaluate agricultural research, today we certainly have a much better set of tools than the ones available two decades ago. Moreover, there are important efforts to operationalise these tools and bring them in the mainstream of research system review and assessment process. There have been many studies and workshops on the subject of impact assessment of international and national agricultural research. Impact assessment has become an important focal area of research in each of the CGIAR centers, particularly since the conduct of the comprehensive study on the impact of the CGIAR system during 1984/85. They have carried out a number of detailed economic evaluations of the returns to research both at the center-level and at the country/regional level using case studies of research impacts. Within the World Bank, impact evaluation of agricultural research projects is receiving increased attention. Although not related directly to agricultural research funding-projects, the recent study by Ezemenari, Rudqvist and Subbarao of the Poverty Reduction and Economic Management Network in the World Bank provides a good overview of the methodology and concepts related to impact evaluation of development projects.

At the NARS level, there are concerted efforts towards practical application of economic evaluation methods as well. Some countries are further ahead in this than others. Australia is perhaps a good example of the most systematic and thorough practical application of economics to research evaluation. Most research institutes and state research departments now require ex ante cost-benefit analysis to be carried out on research proposals before they are funded. To carry out these project evaluations, several organisations have developed spreadsheet and other models to try to simplify and speed up the process. Some organisations have linked their project evaluation software into databases for use in wider priority setting and analysis.

The state governments in developed countries have sought to improve the efficiency and management of public sector by funding several formal economic evaluation studies. These include for example, the evaluation studies in Victoria, Australia funded by the Department of Agriculture, Energy and Minerals, and the evaluation studies in California, USA, by the University of California, Division of Agriculture and Natural Resources.

In the context of developing countries, the progress in the practical application of economic evaluation has been uneven. Research institutes and programs such as KARI in

Kenya, EMBRAPA in Brazil, INTA in Argentina, GRDP in Ghana, PCARRD in the Philippines, ARC in the Republic of South Africa and the national research system in Nigeria are perhaps the pioneers in integrating formal impact assessment within their research systems. In most developing countries, impact assessment studies still remain "externally driven" – conducted through donor funding by social scientists from local or foreign universities. For example, in early 1990s USAID funded US universities such as Michigan State University and Purdue University to conduct several economic impact studies in Africa. Many economic impact studies have also been conducted by developing country nationals as part of collaborative projects with IARCs.

ISNAR has also been instrumental in developing and implementing practical approaches towards ex ante economic assessment and priority setting in the context of developing countries. In recent years, there has been a surge in software packages that apply various benefitcost techniques to agricultural research impact evaluation. A review of these evaluation softwares by Wood distinguishes them into two distinct groups. The first relates to the strategic decisions at the institutional or program level and the second relates to project evaluation. These evaluation tools share several common features. They all are explicitly designed for agricultural research, rely on benefit-cost analysis methods, and are being used for research analysis purposes in various national and international contexts. Wood reports ex post applications of some of the packages. For example, MODEXC and DREAM have been used for ex post analysis by Sanint, and Sanint and Wood. The GIS techniques are also increasingly used in ex ante and ex post impact assessment, which allows better estimates of some of the technical parameters used to assess changes in productivity and supply shifts. With the increasing use of computer softwares and GIS techniques, it is now becoming possible to incorporate socio-economic, policy and environmental variables in an ex ante economic assessment analysis.

Environmental Impact Assessment

Environmental Impact Assessment (EIA) is defined as the process of identifying, predicting, evaluating and mitigating the biophysical, social, and other relevant effects of development proposals prior to major decisions being taken and commitments made. The International Association for Impact Assessment (IAIA) views EIA as a strictly ex ante exercise. The objectives of EIA, according to their recently published guidelines are: (1) to ensure that environmental considerations are explicitly addressed and incorporated into the development decision making process; (2) to anticipate and avoid, minimise or offset the adverse significant biophysical, social and other relevant effects of development proposals; (3) to protect the productivity and capacity of natural systems and the ecological processes which maintain their functions; and (4) to promote development that is sustainable and optimises resource use and management opportunities.

Many countries require environmental impact assessments (EIAs) for major development projects; and, in fact, many countries have formal requirements in law and associated guidelines for carrying out EIAs. At the same time, an OECD study found that few guidelines were implemented in practice because of the lack of human and financial resources, their general, non-specific and (often) mechanistic nature, and their lack of relevance to the main tasks and problems facing guideline users. The experience of the African Development Bank in integrating EIA in their development projects also echoes some of these implementation problems.

The importance of EIA is increasing in agricultural research due to the growing concerns of land degradation, deforestation and loss of biodiversity around the world. The Convention on Biological Diversity negotiated at the Earth Summit in Rio de Janeiro in 1992 requires biodiversity to be considered in impact assessments. Although, this has not been formally adopted as a legislation in CBD member countries, it certainly points to the growing importance of incorporating biodiversity issues in impact assessment of agricultural research. The impact assessment community in general has recently been debating the implications of CBD and impact assessment on biodiversity.

Agriculture's close association with natural and biological resources makes EIA a relevant tool for agricultural research planning. There is a sizable literature on assessing environmental impacts of changes in the rate of natural resource (RNR) use in a general sense. There are also a number of techniques available to place economic values on environmental changes. However, there are few examples of countries and research institutions that have formally assessed the environmental impacts associated with agricultural research. Environmental costs and benefits are typically not included in conventional economic impact studies discussed above. The meta-analysis of the returns to agricultural research done by Alston et al. found that out of more than 1,100 research evaluation observations, only 11 had included environmental variables in the rate of return analysis.

A major criticism of past approaches to impact assessment is that they ignore the fact that the improved technologies produced by agricultural research have, or often may have, favorable and unfavorable consequences for the natural resource base and the environment. Among other things, there is a clear lack of adequate data on which to base EIA. Thus, most of the work that has been done is qualitative and speculative at best. The IAEG is just now starting on this task, looking specifically at research done in the CGIAR centers. An overview of literature on assessing the impact of agricultural research on the environment is included in IAEG's initial report.

Agricultural researchers should review the available documentation on actual and conceptually potential impacts of proposed and past research on the environment and then:

— Describe the potentially expected positive and negative impacts associated with his or her specific research in terms of the expected "with and without" changes in biophysical state of natural resources with adoption of research innovations;

— Attempt to assess the human welfare changes (social and economic) associated with biophysical changes such as decreased risk of flooding loss; reduction of the water table at a slower rate than earlier; changes in deforestation due to creating more sustainable agricultural technologies, etc.

— Provide guidelines on future baseline data requirements and set up M&E systems for tracing environmental impacts in the future.

It should be noted that traditional environmental impact assessments focus on adverse impacts on the environment. Assessments of agricultural research impacts should also include the potential positive or beneficial effects (e.g., social and economic benefits from land saving technologies).

The recent IAEG report referred to earlier on environmental impact assessment of CGIAR productivity enhancing research illustrates some of the issues and the opportunities to evaluate positive impacts of agricultural research. In his foreword, the IAEG Chair assesses the report as follows:

— The review of the available literature on assessment of the impacts of agricultural research on the environment provides important insights, particularly related to the lack of previous empirical work on such impacts.

— The development of a conceptual framework for assessing environmental impacts provides a sound basis for future evaluation of such impacts. Its clarification of the impact pathways from research to environmental impacts is helpful in terms of developing operational approaches to environmental impact assessment.

— The indirect impacts of CGIAR productivity-enhancing research on the environment through the reduction in land area cleared for agriculture are significant.

Social Impact Assessment

Social impact assessment (SIA) is described as the process of identifying the future consequences of a current or a proposed action, which are related to individuals, organisations and social macro-systems. According to this definition, social impacts are the consequences for human populations of any public or private action that alters the way in which people live, work, play, relate to one another, organise to meet their needs, and generally cope as members of society. There are also links between SIA and economic impact assessment, which is concerned with the impact of actions on economic structures and processes, including changes in employment, business activities, earnings and income.

Central to SIA is the ex ante assessment of consequences of policies/programs/ initiatives/actions on social, institutional and individual behavior. In this sense, SIA is relevant for agricultural research and technology development, since they aim to have an impact on social and human behavior. This, for example, was made clear in the IAEG-sponsored study by Sechrest et al. into adoption of technological innovations.

SIA was formalised in early 1970s with its incorporation in EIA that had become mandatory with the introduction of the U.S. National Environmental Policy Act of1969. Today SIA is an important component of activities of the International Association for Impact Assessment (IAIA). In 1986, the World Bank decided to include both environmental and social assessment in their project evaluation procedures because liabilities were increasing for projects evaluated strictly on economic and financial criteria. In the US, SIA reached its highest legitimacy when at the conclusion of the April 1993 'Forest Summit' in Portland, Oregon, President Clinton mandated that a social assessment of each timber-dependent community would be a required component in deciding among alternative management futures for old growth forests in Pacific Northwest.

Over the years, SIA has evolved as a distinct field under impact assessment. While SIA is normally undertaken within the relevant national environmental policy framework, it is not restricted to this, and SIA as a process and methodology has the potential to contribute greatly to the planning process of other types of development projects. For agricultural research, it can assist in the process of evaluation of alternatives, and to help in their understanding and management of the process of social change. However, based on a review of available studies, it is evident that SIA has rarely been applied to agricultural research programs. The estimates of social surplus in impact studies of agricultural research are based on costs and benefits that are measurable in monetary units. The social surplus methodology used in economic assessment is amenable to estimating distributional consequences of research, such as between consumers and producers, and between different income groups of consumers and producers. But apart from this partitioning of benefits, the closest, economic studies have gone in including social impacts is through qualitative assessments. In ex ante evaluations, social impact indicators (such as gender impacts) are sometimes included as one of the criteria in scoring models.

Social impacts are important and need to be considered along with the economic and environmental impacts. However, conducting the economic, social and environmental impact assessment of a research program as separate disciplinary activities may be too burdensome for research managers in developing countries. The challenge is to focus on some specific social, economic and environmental issues and then to explore the most appropriate methods to address them. A research organisation should foster a multi-disciplinary approach to impact assessment so as to use the strengths of various disciplines in the impact analysis. For example, economic assessment can better deal with attribution problems than the social and environmental impact analysis. On the other hand, the latter are able to better deal with analysing the constraints to adoption and factors affecting technology impacts that are less amenable to quantification. This will enrich the impact analysis as well as provide a clearer identification of issues for research planning and prioritisation.

Institutional Impact Assessment

Increasing agricultural productivity, whilst strengthening local institutions, has long been an important goal of agricultural research. Agricultural research organisations play an important role in meeting this goal by improving technologies and knowledge base of the biological, social, economic and political factors that govern the performance of an agricultural system, and by strengthening local institutions' capacity and performance. While research projects themselves are often subjected to rigorous appraisals from an economic, social and environmental perspectives, research methods and institutional aspects of a research organisation tend to escape any kind of impact analysis. The impact assessment work discussed above focuses on the impact evaluation of the "technological" outputs of research organisations in the form of new techniques, methods, information and practices of agricultural systems. Institutional impact assessment involves the evaluation of the performance of a research organisation in non-technological research activities such as training, networking, development of methodologies, and advisory services in the areas of research and other policies, organisation and management. Many national and international agricultural research institutes devote sizeable resources on such activities. Assessment of the institutional impacts of such activities should therefore be an integral part of the overall impact assessment and research evaluation efforts.

There has been little methodological and practical work in the area of institutional impact assessment of agricultural research. This includes the impact an agricultural research organisation has on capacity building, human resources development, and performance of other institutions. However, the recent surge in the interest to evaluate the "institutional impacts" of CGIAR centers, has led to significant advances in the methods and application of "institutional impact assessment" by centers such as ISNAR and IFPRI.

ISNAR, for example, has undertaken a major effort in this area and generated several studies and results that illustrate the conceptual and analytical methods of institutional impact assessment. These studies conceptualise the institutional impact of an organisation (in this case ISNAR) on the environment, motivation and capacity of a client organisation (in this case a national agricultural research organisation). The institutional performance and impact of ISNAR was measured by undertaking five component studies that included: a survey of NARS leaders, a survey of ISNAR's key stakeholders, three country case studies, a metaevaluation of ISNAR review reports and an inventory of ISNAR outputs.

IFPRI has also recently undertaken several case studies to document the institutional impacts of its policy research and capacity building activities. Determining how policy research, and strengthening of the analytical and decision-making capacity affect policy choices is not easy. Nonetheless, IFPRI's experience and the use of case study methodology shows that it is possible to discern links between institutional capacity building (including policy research) and policy actions.

The CGIAR as a system also on a regular basis evaluates its 16 centres through its external program and management review process, using outside evaluation expertise. These evaluations tend to be forward looking and comprehensive. Part of the exercise involves specifically looking at the actual and potential impacts of the centres' activities. The IAEG is initiating such work related to the impacts of formal and informal training in the CGIAR and its impacts on scientific and institutional research capacity in national agricultural research systems. Previous work by ISNAR and other groups will be drawn upon, including work recently reported in the Winter 1999 volume of Knowledge, Technology & Policy which is devoted to "evaluation in developing countries."

The concrete results and impacts of institutional development activities of a research organisation can be difficult to see and may take time to emerge. However, information generated from institutional impact assessment has the great potential to lead to better, more effective actions and institutional performance of a research system.

IMPLEMENTATION AND INSTITUTIONALISATION OF IMPACT ASSESSMENT

The fields of economic, social and environmental impact assessments (in the general sense) have made significant strides in the state of the art in methodological front. In the context of agricultural research, the methods for economic impact studies have progressed much more both in theory and practice, than the social and environmental impact assessment methods. However, to be effective as a tool for research planning and efficient resource allocation, impact assessment (whether it is economic, social or environmental) needs to become an "internally-driven" exercise within the research system. The important question facing us today is how to use these available methods to implement agricultural research impact studies in developing countries. Below we discuss some institutional questions that need to be addressed in the implementation of impact studies at a country level.

Attributing Development Impacts to Research

An important issue from the point of view of the impact evaluator is how to deal with the attribution problem. In order to be accountable, one should be able to attribute results (impacts) to the activities on which resources are invested. Yet, in the case of research, it is a long ways from investment of resources in the process through to results and ultimately their use in generating measurable impacts in terms of poverty reduction, food security, or environmental improvement. There are many groups/factors that enter the picture along the way between research and impact in terms of ultimate development goals.

For example, in an area where maize yields or milk production have increased, how can the contribution of research versus that of extension, credit programs, and improvements in market conditions be estimated? All of them would like to, and probably can claim some share of the credit for the impacts (or blame if impacts are negative) ultimately deriving from the research. The question from an institutional perspective is how to isolate and attribute impact value specifically to the research component when other institutions (e.g., extension service, regional and international research organisations), and policies and programs (e.g., credit system, input subsidies, marketing infrastructure) have contributed to the development goals? This is basically a conceptual problem that can potentially hinder the implementation and institutionalisation of impact assessment. A number of approaches have been used in economic assessment, ranging from arbitrary allocation of total benefits among the various actors involved based on the share of total costs expended, to including the costs of all the contributing agencies in estimating a total rate of return to investments, to making no attempt at all at partitioning the impact by institutions.

Attribution problem is especially difficult to deal with in social, environmental and institutional impact assessments. The problem is severe as the unit of analysis of IA becomes smaller (such as individual project-level). Kuby as well as in the lead paper for this workshop argues for the separation of monitoring responsibilities with projects monitoring inputs, outputs and outcomes, and independent local groups charged with monitoring social impacts. It is true that the latter responsibility is beyond the capacity of the research project itself. However, at an aggregate research program or a research system level, efforts should be made to measure the impacts of research as they affect the broad social and environmental goals.

Importance of Impact Assessment Studies

Most of the impact assessment studies in developing countries found in the literature have been done as an academic research study leading towards a post-graduate degree in a local or a foreign University. This type of research studies conducted independently by researchers and universities is important to ensure quality and to generate pressures within the country to address problems identified by these assessments. These studies have been, in most cases funded, partially or fully, by donor agencies. It is not clear, however, whether such impact studies are integrated by a developing country NARS and research programs in their research evaluation and planning system.

An important question therefore is how to increase the impact of impact studies. Looking at the cases of industrialised countries such as Australia and developing countries such as Brazil, Argentina, Kenya, and Philippines (who all have research evaluation units within their research system), it is evident that the impact of IA will depend on whether or not NARIs build an "in-house" capacity to conduct IA, and how successful it is in fostering a self-evaluative culture within the research system. Byerlee and Alex provide some guidelines for institutionalising impact assessment in national research institutes, through the establishment of monitoring and evaluation capacities.

Research institutes need to have an "impact assessment" unit (may be as part of an overall research evaluation unit) or a champion responsible to provide leadership in conducting IA and integrating their results in the research system planning and resource allocation decision making process. The establishment of a Unit for Development Impact Analysis (DIA) by the Agricultural Research Council (ARC) in South Africa in 1995 is a good example of such a unit. The DIA is responsible for planning and prioritisation of research within the ARC, and to mobilise additional resources for impact assessment.

An alternative for research systems that do not have socioeconomic research capacity to conduct impact assessment is to contract out such studies to independent agencies and universities and broaden their institutional linkages with other research institutes within a region. The important thing, however, is that the "ownership" of impact studies should remain with the country or research institutes whose research programs are being assessed. The involvement of decision-makers and researchers of the research institutes is the key to the successful use of the results of such impact studies. If IA is to become an effective learning tool for institution building, it is important for a research institute to maintain objectivity in the whole process and to make the results of the assessment public.

Type and Intensity of Impact Assessment

The question often raised is whether there is one "right" type of IA that can be recommended for agricultural research? The answer is "No." So what determines the type, focus and intensity of IA chosen out of the available portfolio of methods and approaches?

Figure 3 sums up the main factors determining the type, intensity and focus of an IA undertaken for agricultural research. The choice will depend on the purpose and objectives, the types of impact dimensions of interest to the decision-maker, the resources available, and the previous knowledge of and experience with IA (perceptions of recipients regarding credibility, reliability, etc.).

The techniques of IA discussed above are used for different purposes and under different circumstances. Most commonly, the favored assessment is one that focuses on the value of returns on resources invested. Such estimates are useful input for making decisions on allocations of resources to research vs. other uses for the resources. They also are useful in politically justifying past expenditures of resources.

Institutionalising Data Systems for Impact Assessment

Estimating impacts of agricultural research is complicated

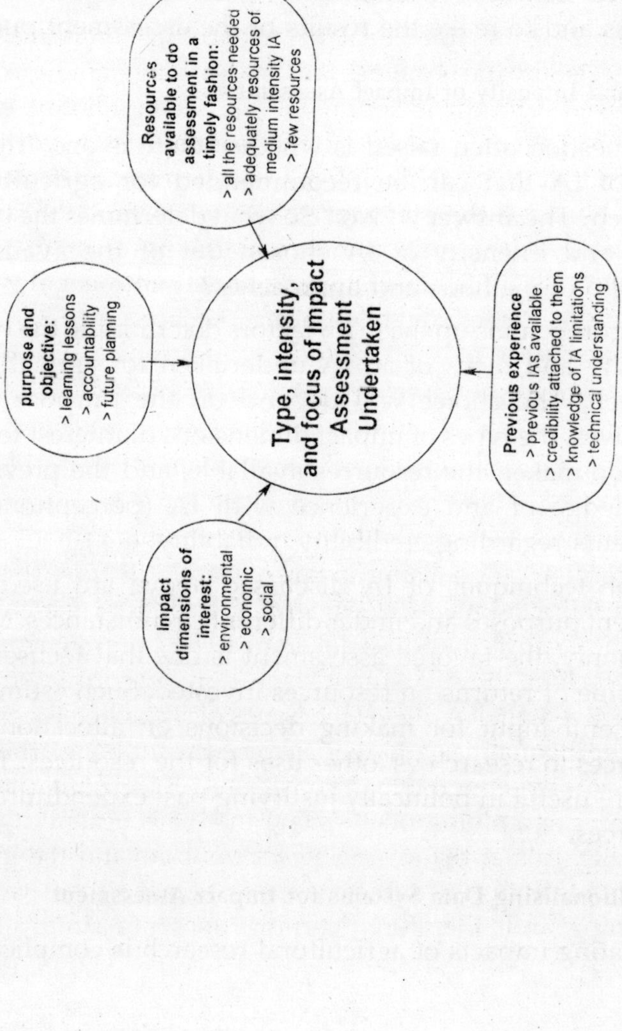

Figure 3. Factors Determining the Type, Intensity and Focus of Impact Assessment

by the dispersed area over which impacts may occur and the large number of external factors affecting target indicators. The full set of indicators needed for a typical IA study is large, including the broad goal-level impacts (quality-of-life indicators, farmer incomes, environmental conditions) resulting from research outcomes (productivity change, technology adoption, institutional development) and the program activity measurements (input and outputs) that lead to these changes. IA is a dataintensive activity and the limited progress in SIA and EIA in agricultural research is to some extent because of data limitations. Research programs must balance data needs between the ideal (and costly) and the practical, and must draw data from various sources. These include: data from internal management information systems such as experimental data, regular adoption surveys in defined areas, special studies using techniques such as surveys, rapid rural appraisal, key informant interviews, focus group interviews, community interviews, structured direct observations, and informal surveys, special topic studies, and diagnostic case studies. National databases are another sources to measure changes in goal-level indicators on a national basis. These are government-sanctioned statistics that are readily available and comprehensive, representing considerable investment in data collection.

Cost of Impact Assessment

Assessing the effects of agricultural research is complex and costly for several reasons. First, it requires costly fieldwork and analysis. Building and institutionalising the data systems for IA also requires organisational and financial resources. Second, it requires expertise and human resources to analyse data and assess the impacts. There is no general guideline on how much to spend on impact assessment. The cost will depend on the scale of impact assessment (national-, program- or project-level) and depth of assessment (methods used and rigor applied). IAs can be low cost involving one researcher observing what appears to have happened or is happening due to a project or activity in terms of meeting goals. Or they can be highly sophisticated and often expensive formal evaluations using recognised outside expertise, developing high levels of detail with verifiable levels of accuracy and reliability.

The economics of impact assessment itself is not well documented or evaluated in the literature. To institutionalise impact assessment requires the development of guidelines on budget share for impact assessment. As research systems "mature" and internalise the impact assessment process, they should adopt some "rules of thumb" such as a certain percentage of total research budget to be allocated for impact assessment. For example, GTS aims to spend on project evaluation (including all self- and independent evaluation activities) not more than 5% of a research program's budget on average. One of the summary discussions of a research evaluation workshop organised by GRDC in Australia was that 3-4% of the research and development budget could be spent profitably on impact evaluation.

IMPLICATIONS FOR THE FUTURE

Recognising and Adapting to Limitations and Constraints

The fundamental nature of the research process and the probability of significant impacts needs to be understood by all, including the awareness that all research does not produce major "breakthroughs." Many research activities quite understandably lead to "dead ends" that have no impact in terms of goals. It also needs to be understood that most research will neither be a breakthrough nor a dead end, because it will not be possible to assess it's impact. Walker and Crissman, in justifying their selection of success stories for impact evaluation, contend that,

> The belief that every agricultural research project will successfully generate a practical impact is decidedly mistaken, but the expectation that a mature agricultural research program will have documented one or more success stories is quite reasonable.

In research we have to be comfortable with the idea that it is total investment that needs to pay off in terms of goals. Thus, impact assessment should be carried out at a unit of account greater than a research project (such as the research program or a research institute). Documenting key stories of success and failure is essential to arrive at an initial understanding of the productivity of an agricultural research program, and to convince the donors and research sponsors of the economic value of their investment. The benefit of studying success stories is illustrated by the examples of CIP-related technologies and ARC in South Africa.

There needs to be an awareness of the limitations on trying to attribute development impacts back to research. This awareness is needed on the part of both funders and researchers, as well as research administrators. Similarly, research managers and administrators need to be aware of the limitations of the means or methods available to evaluate impacts at different levels of research (strategic, applied and adaptive). While many good techniques are available, it is important that potential demanders and recipients of impact assessments understand the data limitations that exist in most cases. This limitation was fully accepted by the impact assessment team in the case of Agriculture Victoria. Out of the 40 research projects evaluated, 18 were evaluated quantitatively using cost-benefit analysis, while for the other 22 projects only qualitative assessments were done due to the nature of the research project and data available (e.g., policy research projects).

Another related constraint that needs to be recognised is that resources to conduct impact assessment are limited in all cases, and this may affect the quality and quantity of impact studies. The important first step is to initiate impact assessment by undertaking one or two studies and utilising a simple analytical framework to document research benefits, even in situations where data are scarce. Such an exercise will focus attention of managers and researchers on evaluation in general, which will provide a useful starting point for in-country discussions on research impact and assessment and mobilise resources to conduct further assessments.

Research managers and policymakers tend to be skeptical of the data and methods used in impact assessment; they may also find the reports difficult to understand, interpret and apply. This highlights the need to plan impact studies in terms of real information needs (rather than peer interests), to pay close attention to data quality, and to make special efforts to summarise the findings. It is extremely important for results and recommendations to be presented in terms that are meaningful to policymakers, managers, and scientists.

Practical Guidelines

The impact mentality, culture and awareness need to be instilled in those who work in the organisation. Researchers should be aware and fully support the idea that they need to be accountable for investment of resources, whether or not such investment results in positive impacts in terms of the goals of the organisation or program involved. Researchers also need support in using IA in a dynamic sense in the planning of future activities and programs.

Many types of impacts can only be adequately assessed if relevant data exist, and such data can only be obtained if baselines are established and monitoring of changes takes place. In the language of economists, the application of a "before and after" scenario of a given research innovation is not likely to be the same as the situation "with and without" the innovation. Changes would have occurred even without the innovation. We need to be monitoring changes as they occur.

To sustain and implement the impact assessment work on a regular basis, a share of an organisation's budget should be dedicated to IA activity, including the setting up of an M&E system to trace ongoing and emerging results from current programs. In most research organisations, it will be important that at least one identifiable staff member focuses on impact assessment and helps to develop the impact culture of the organisation. This person will also most likely have major responsibility for providing liaison with funders and decision makers, learning what their impact assessment needs are and, at the same time, making them aware of what can and cannot be expected from assessments of impacts of various kinds and under various budget constraints. Judicious use of outside expertise may also be called for, but this is no substitute for having in-house capacity and an impact mentality in the organisation. There are several ways of inculcating the impact culture and mentality in a research organisation:

1. Make researchers focus on the value side of their work as well as its technical merit.
2. Increase the interaction between scientists and impact evaluators at the initiation and development stages of program proposals.
3. Require all new proposals above some cut-off budget value to be accompanied by an ex ante impact assessment.
4. Make M&E of ongoing research a routine practice, requiring research programs to submit progress reports, achievements of milestones, final reports and undergo technical and financial audits.

5. Make IA an ongoing activity by emphasizing the importance of implementing ex ante assessment, M&E and ex post assessment at different levels of research aggregation.

REFERENCES

Echeverria, R.G., "Assessing the impact of agricultural research", In R. Echeverria (ed.) *Methods for diagnosing research system constraints and assessing the impact of agricultural research*, Vol. II, Assessing the impact of agricultural research, The Hague: ISNAR, 1990.

Elbasha, E., P.K. Thornton, and G. Tarawali, "An ex post economic impact assessment of planted forages in West Africa", *LRI Impact Assessment Series 2*, Nairobi, Kenya: ILRI, 1999.

Evenson, R.E., P.E. Waggoner, and V.W. Ruttan, "Economic benefits from research: An example from agriculture" *Science*, 1979.

Oehmke, J. and E. Crawford, "The impact of agricultural technology in subsaharan Africa,""*Journal of African Economies*, 1996.

Omiti, J., M. Waithaka, and W. Mwangi, "Situation of impact assessment of agricultural research in the ASARECA region", *Paper prepared for the workshop on impact assessment of agricultural research in the East and Central Africa*, Entebbe, Uganda, November 16-19,1999.

Ozgediz, S., "Evaluating research institutions: lessons from the CGIAR", In *Knowledge, Technology and Policy*. Vol. 11, 1999.

Paarlberg, R.L., "External impact assessment of IFPRI's 2020 Vision for food, agriculture, and the environment initiative", *IFPRI Impact Assessment Discussion Paper 10*, Washington, D.C.: IFPRI, 1999.

11

Functions of Indian Institute of Horticultural Research

The Indian Institute of Horticultural Research (IIHR) was established by the Indian Council of Agricultural Research (ICAR) towards the end of fourth five-year plan. It started operating in New Delhi on 5 September 1967 and was subsequently shifted to Bangalore on 1 February 1968 at Hessaraghatta where a Fruit Research Station was started as early as February 1938 by the Imperial Council of Agricultural Research, which in due course was converted into National Hortorium, located 26 km north of Bangalore on the Bangalore - Bombay line. IIHR was the first of its kind in the country mandated specifically to horticultural crops. Initially, 24.7 ha land under the National Hortorium was transferred and later on Government of Karnataka gave 238 ha. land in the village Ivarakandapura opposite the National Hortorium. The Institute started with a skeletal staff with 7 research divisions manned by 50 scientific and technical staff in various disciplines. Over the years, the Institute has grown up in its size and sphere of activities and now is one of the pre-eminent horticultural research institutes in Asia. Today, the Institute has completed 28 years of service to horticultural research and development.

MANDATE

The Institute is well set to work with the following mandate:

1. To undertake basic and applied research for developing strategies to enhance productivity and utilisation of tropical horticultural crops, viz., fruits, vegetables, ornamentals, medicinal and aromatic plants and mushrooms.
2. To act as a repository of scientific information relevant to horticulture.
3. To act as a centre for training for upgrading of scientific manpower in modern technologies for horticultural production.
4. To collaborate with relevant national and international agencies in achieving the above objectives.

SET-UP AND ACTIVITIES

The main experimental farm established at Bangalore has a total area of 266 ha which is divided into 9 blocks working on various horticultural crops, viz., fruits, vegetables, ornamental, and medicinal and aromatic crops including mushrooms with 10 full fledged divisions and 4 sections apart from two units of All India Coordinated Projects on Tropical Fruits and Betelvine with well equipped laboratories. In addition, the Institute is also involved in All India Coordinated Research work on specific problems carried out with international collaborations and bilateral agreements with other countries. The Institute is also involved in Post-graduate education and training in collaboration with the University of Agricultural Sciences, Bangalore, and Dharwad and Birsa Agricultural University, Ranchi. Many of the scientists are recognised as guides for the University students.

Presently, the Institute has got four experimental stations, viz., Central Horticultural Experiment Station, Chettalli; Central Horticultural Experiment Station, Godhra; Central Horticultural Experiment Station, Ranchi and Central Horticul-tural Experiment Station,

Bhubaneshwar which take care of research needs of various states and adjoining areas. The research needs of tribal belts of eastern and western region of the country are also taken care of by these centres and Krishi Vigyan Kendra at Gonikoppal, Karnataka. The Institute is headed by a Director, supported by scientific and technical staff, administration, accounts and other supporting staff, which now number over 700.

SIGNIFICANT ACHIEVEMENTS

Crop Improvement

Germplasm collecting, maintenance and conservation

Field germplasm collections: Valuable germplasm of fruits, vegetables, ornamentals, medicinal and aromatic crops have been collected within the country as well as introduced from abroad through the National Bureau of Plant Genetic Resources, New Delhi. The number of collections presently available at the Institute are - fruits: 1669, vegetables: 5509, ornamentals: 1092, and medicinal and aromatic crops: 80. (Table 1) These germplasms are being maintained in field by the breeders and are used for breeding programmes.

Genebank: The Institute has established a genebank to conserve horticultural germplasm for short, medium and long term periods. It has a number of cold chambers with temperature varying from +25^{0}C to -20^{0}C. The seed is processed, packed in suitable containers and stored at temperatures according to the requirement of each crop. The genetic information as well as placement of germplasm in racks in cold chambers have been computerised for quick display and retrieval. The IIHR genebank has a capacity to hold more than 200 000 accessions, so far about 2500 accessions have been stored (Table 2).

Table 1: Germplasm of horticultural crops available at IIHR

Name of the crop		*No. of accessions*
	Fruits	
1.	Mango	170
2.	Banana	290
3.	Grape	579
4.	Citrus rootstocks	12
5.	Limes and lemons	70
6.	Guava	24
7.	Pomegranate	52
8.	Fig	16
9.	Jujube	38
10.	Apple	3
11.	Kiwi	3
12.	Others	9
13.	Watermelon	135
14.	Muskmelon	268
	Sub-total	1 669
Vegetables		
1.	Amaranthus	140
2.	Bitter gourd	40
3.	Cabbage	425
4.	Bottle gourd	106
5.	Brinjal	220
6.	Broccoli	25
7.	Chillies	1 000
8.	Carrot	298
9.	Cauliflower	215
10.	Chinese cabbage	32
11.	Cowpea	50
12.	Cucumber	158
13.	Dolichos	15
14.	French bean	180
15.	Okra	125

16.	Onion	601
17.	Palak/spinach	15
18.	Sponge gourd	89
19.	Pumpkin	35
20.	Ridge gourd	40
21.	Sweet pepper	200
22.	Tomato	1 200
23.	Garden peas	300
	Sub-total	5 509
Ornamentals		
1.	Bougainvillea	119
2.	Roses	464
3.	Gladiolus	120
4.	Chrysanthemum	180
5.	China aster	21
6.	Hippeastrum	52
7.	Hemerocallis	34
8.	Orchids	102
	Sub-total	1 092
Medicinal and Aromatic Crops		
1.	Catharanthus roseus	11
2.	Dioscorea	6
3.	Jasmine	39
4.	Patchouli	6
5.	Pogostemon	6
6.	Scented geranium	2
7.	Solanum spp.	10
	Sub-total	80
	Grand total	8 350

Seed storage: Research has been in progress to enhance seed longevity in fruits, vegetables and ornamentals by manipulating pre-harvest factors, moisture content, temperature, oxygen etc. Studies have also been carried out on genetic stability, change in physiological and biochemical parameters in stored seed germplasm.

Table 2. Cropwise number of accessions stored in genebank at -20°C

Sl.No.	Name of crop	No.of accessions
Fruits		
1.	Citrus	101
2.	Papaya	16
3.	Muskmelon	3
	Sub-total	120
Vegetables		
1.	Brinjal	115
2.	Chilli	1,317
3.	Cluster bean	26
4.	Cucumber	2
5.	China aster	12
6.	Garden peas	45
7.	Okra	68
8.	Ridgegourd	45
9.	Solanum sp.	2
10.	Tomato	744
	Sub-total	**2 376**
	Grand total	**2 496**

Among fruits, seeds of papaya, lemon, kagzi lime, mandarin, citrus rootstocks, ber/jujube, mango and phalsa have been studied for their storage qualities. Papaya seeds were found viable for short periods (6 months) under ambient conditions, but their longevity was extended to four years when stored with silica gel, which increased further by storage at 10°C. Leaching of soluble sugars and free amino acids was found to increase and dehydrogenase activity decreased due to ageing of seeds.

Seed viability of kagzi lime was retained for 5 years when stored in polyethylene and aluminium foil laminated bags at 5°C or -18°C with imbibed storage at 5°C. Storage of fruits at 5°C or - 15°C was also found to preserve viability for 120 days. Seed drying was found to reduce germination of seeds of different citrus rootstocks. In Coorg mandarin, seed was found to lose viability rapidly under ambient conditions, and none germinated after 30 days. This was accompanied by increased leaching of soluble sugars, amino acids and decreased dehydrogenase activity in seeds. In ber, the seed viability was lost rapidly under ambient conditions while it was retained for 4 years at 5^0C and -18^0C. Seeds were killed by freezing when stored in paper bags at $- 18^0C$. Mango seeds exhibited recalcitrant storage behaviour. In cvs. 'Alphonso' and 'Dashehari', seed longevity was extended from 15 days to 365 days by imbibed storage at 15^0C. Phalsa seeds remained viable for 5 years at 5^0C and -18^0C. Seeds of grape cv. 'Bangalore Blue' could be preserved for 2 years at 5^0C and $- 18^0C$. Dormancy could be eliminated by treating seeds with gibberellic acid (2000 ppm).

Vegetable seeds exhibit orthodox storage behaviour and are easy to conserve. Genetic differences for seed longevity were observed in onion, chilli, cauliflower and brinjal, and classified as short, moderate and long term for storage under ambient conditions. Seed maturity which affects the storage life varied from several days to weeks in different vegetable crops. Seeds from immature fruits did not germinate in chilli cvs. Pant C-1 and Arka Lohit. Seed longevity and vigour were higher in seeds harvested at ripe to overripe stages which coincided with the presence of high quantity of total proteins and free amino acids in seeds. Drying at higher temperature (40^0C) damaged the seeds and resulted in low percentage of germination in okra. High percentage of germination was obtained in chilli when fruits were stored in cloth, paper and polyethylene bags. The incidence of insect attack was more in polyethylene bag-stored fruits.

Effect of antioxidants and inhibitors on storage life of seeds: Effect of chemical treatment with certain antioxidants and inhibitors has been studied in seeds of radish, onion, French bean and cowpea. In radish, 2, 4-dinitrophenol at 0.08%

significantly reduced germination while other chemicals like EDTA, borax, potassium metabisulphate (KMS), oxalic acid, ascorbic acid, abscisic acid (ABA), maleic hydrazide (MH), glucose and polyethylene glycol (PEG) preserved the seed viability. Germination of French bean seeds was affected by all the chemicals except glucose. Abscic acid (ABA) and maleic hydrazide (MH) treatment in onion significantly reduced the germination. Similarly, seed germinability was affected by all the chemicals in cowpea.

Studies on seed storage under controlled temperatures showed that seed could be stored for 10 years in onion, radish, French bean and cluster bean, 8 years in bell pepper and watermelon, 7 years in okra and 4 years in brinjal, tomato, long melon, amaranthus, bottlegourd and pumpkin by storing seeds at 5^0C and - 18^0C. Polyethylene aluminium foil laminated bags and glass containers gave better results than paper containers, especially at sub-zero temperatures, for seed storage.

It was observed when seeds of bell pepper, chilli, lablab bean, okra, onion and watermelon were packed in aluminium foil laminated bags under partial vacuum and stored under ambient conditions, the percentage of germination and seedling vigour were greater. The percentage of germination was higher in nitrogen atmosphere than in carbon dioxide in muskmelon, okra and onion, while with bell pepper it was high with cabron dioxide. Seedling vigour was not affected.

Studies on physiological and biochemical changes in stored seeds indicated that the electrical conductivity, soluble sugars and free amino acid contents were greater in leachates of ambient stored seeds than the low temperature stored seeds. Similarly, dehydrogenase activity decreased in seeds stored under ambient conditions, accompanied by loss of seed germinability and seedling vigour.

Preliminary studies indicated possibilities of conserving orthodox seeds in suitable liquid embedding medium for long term at ambient temperature avoiding refrigeration. Pollen storage: Studies in papaya and citrus have shown that pollen could be kept viable for durations ranging from 2 to 6 months in domestic refrigerators at temperature ranging from 5^0C to - 18^0C. With the establishment of constant temperature regime, pollen under low temperature conditions could be kept viable beyond 6 to 8 months at - 18^0C to - 20^0C, but with a gradual decline in viability tested under in vitro conditions. Freeze dried pollen stored at - 20^0C retained viability and fertility upto 14 months of storage in papaya. Uniform low temperature (either at + 5^0C or at - 20^0C) was found to be very critical for retaining good viability for short and medium term storage. The decline in viability is more rapid at 5^0C than at - 20^0C in pollen of different cultivators. Pollen samples have been stored at - 20^0C in the genebank since 1992. Field pollination's using low temperature-stored pollen of papaya, citrus, grape, tomato, brinjal, rose and gladiolus resulted in satisfactory fruit and seed set with reduced frequency as compared to control. Seeds set through stored pollen germinated rather slowly as compared to seeds obtained through open pollination, indicating loss of seed vigour. Mortality rates were quite substantial at seedling storage, which impeded complete establishment of plants.

More than 400 pollen samples of 26 different species belonging to fruits, vegetables and ornamental crops have been successfully cryopreserved in liquid nitrogen for ten years in 5 cryogenic containers. It was observed that in almost all the crop species, a good measure of viability and fertility could be retained by storage in liquid nitrogen. In fruits such as papaya and grape, pollen have been successfully cryopreserved for 7 years retaining good viability which induced fruit and seed set when used in hybridisation programmes. Citrus and mango pollen have been kept viable under cryogenic conditions for durations ranging from 1 to 4 years and pollen fertility checks were successful after cryostorage.

Among vegetable crops, tomato and egg plant pollen have been cryopreserved for durations as long as 8 to 9 years, respectively. Cryopreserved pollen responded to fertility tests carried out through field pollination exceedingly well, producing seeds similar to those from fresh pollen. Onion pollen cryopreserved for 6 years emerged successfully when tested for fertility under field conditions. Capsicum pollen kept for a duration of 2.5 years in liquid nitrogen was viable in vitro and fertile. Rose pollen was cryopreserved successfully for one year. In all crops tested, no decline in viability was observed. Development of viability assays for stored pollen: The pollen storage laboratory has optimised a number of procedures for testing pollen viability in vitro.

i) In vitro germination in liquid media-modified hanging crop technique,germination in liquid media-modified cellophane technique;
ii) In vitro germination of pollen tubes in stellar material;
iii) Influence of magnetic stimulus to accelerate germination rates in vitro;
iv) Instant pollen tube formation in an acidic media; and
v) Use of fluorescent dyes to index viability.

Results of the viability tests indicated that in general there is a very strong direct correlation between pollen germination and seed set, in some crops, and in some cases an inverse correlation was found between germination and seed set. In crops like papaya, citrus (trifoliate orange), banana and tomato, breeders have utilised the pollen storage, protocols developed at IIHR in their breeding programmes.

VARIETIES RELEASED

Over the years, IIHR has released a number of varieties/ hybrids in fruits, vegetables, ornamental, medicinal and aromatic crops.

Fruit Crops

Mango

Four hybrids have been released in mango so far. They are 'Arka Aruna (Banganapalli × Alphonso), 'Arka Puneet' ((Alphonso × Banganapalli), 'Arka Anmol' (Alphonso × Janardan Pasand) and 'Arka Neelkiran' (Alphonso × Neelum). All the four hybrids are regular bearers and do not have spongy tissue problems. Arka Puneet has excellent colour with red blust and Arka Anmol has excellent keeping quality. Both these have export market. Arka Neelkiran is a late crop.

Grapes

Four grape hybrids, viz., Arkavati, Arka Kanchan, Arka Shyam and Arka Hans have been released after a decade of work on breeding prior to 1985. During 1992, seven superior grape hybrids, four of them "or table purpose, one for juice and two for wine were identified for release. Among these, Arka Neelamani is a black seedless with crisp pulp.

Passion fruit

A promising hybrid between purple and yellow-fruited varieties has been evolved at Chettalli. The hybrid is high yielding and tolerant to leaf spot, root rot and resistant to root knot nematodes.

Papaya

A variety by name Coorg Honey Dew was evolved at

Chettalli Regional Station and an improved papaya selection Pink Flesh Sweet was released by IIHR, Hessaraghatta.

Guava

Two soft seeded guava varieties Arka Mridula and Arka Amulya have been released.

Annona

One interspecific hybrid in annona named as Arka Sahan has been released which has good keeping quality and high TSS.

Litchi

One selection in litchi has been identified and released from CHES, Ranchi under the name Swarna Roopa which is resistant to fruit cracking with high TSS. This is the first litchi variety released in the country.

Jujube/Ber

A promising ber selection CHES-1 from variety Umran has been released which is earlier than Umran, high yielder and escapes insect, pest disease problems and has been christened as Gomati Kirti.

Pomegranate

Two pomegranate hybrid with dark red arils and soft seeded superior to Ganesh seedless have been developed and are in the advanced stages of testing.

Vegetable Crops

The Institute has so far released 57 improved vegetable varieties of which 31 varieties have been released at the national level and 26 varieties have been released at the state level. Major breakthrough has been made by developing okra varieties Arka Anamika and Arka Abhay resistant to yellow vein mosaic virus, tomato F1 hybrids Arka Abhijit and Arka Shreshta resistant to bacterial wilt, onion variety, Arka Kalyan tolerant to purple blotch and muskmelon variety, Arka Rajhan resistant to powdery mildew. Breakthrough has also been made in evolving multiple disease resistant varieties like Arka Manik watermelon with triple disease resistance to anthracnose, powdery mildew and downy mildew.

Ornamental Crops

The Institute has released 74 varieties in various ornamental crops, viz., gladiolus, bougainvillea, hibiscus, China aster, chrysanthemum, rose, tuberose, croton, etc. These have been bred for various traits like colour, loose flower, display and bedding purposes and resistance to pest and diseases. Protocols have been developed for in vitro propagation of orchids and anthuriums.

Medicinal and Aromatic Crops

In medicinal crops, first breakthrough was made in 1974 by developing two varieties in Dioscorea, viz., Arka Upkar and FBI(C)/1 having high diosgenin content. Similarly, Arka Sanjeevini a less spiny variety and Arka Mahima a tetraploid with high solasodine content in Solanum viarum have been developed and are being commercially grown for production of steroids for family planning purposes. In aromatic crops, one Jasmine Selection Arka Surabhi has been identified for release which has high oil content.

CROP PRODUCTION

Fruit Crops

Mango

In mango, contribution of tree age towards yield is

important and fertiliser scheduling has been made age specific. Vellaikolumban rootstock imparted dwarfing in Alphonso scion and tree volume was reduced by 40 per cent with higher productivity. Shedding of fruits in mango beyond marble stage of growth was directly related to canopy size. Surface temperature was less in affected fruits as compared to fruits with spongy tissues.This could be made use to screen the fruits non-destructively. Paclobutrazol treated trees showed lesser total cytokinin activity and hence its possible dwarfing effect.

Guava

For Sardar guava, optimum fertiliser dose for higher yield was 900:600:600 g NPK/ tree. But for increasing the fruit quality, the N should be reduced by 33%.

Papaya

Papaya with 2262 plants/ha and six split applications of NPK (250:250:500g N, P_2O_5, K_2O/plant/year) was the best for Coorg Honey Dew Cultivar.

Banana

High density planting in banana (4444 plants/ha) resulted in 33% higher yields. Drip irrigation in banana could economise the water by 25% and with a yield increase of 10%, fertiliser with nitrogen reduced the requirement by nearly 50%. Effective weed control could be achieved in banana using diuron at the rate of 2 kg a.i./ha.

Grapes

Thompson Seedless and Arkavati grapes raised on 1613 rootstock gave highest yields. The best training system of yield and quality in Thompson Seedless grape was 8-arm Tatura and 12-arm Tatura for Arkavati grapes. Use of cycocel twice (1000 ppm) after back pruning and smearing with hydrogen cynamid (1.5%) after October pruning increased the cluster/cane ratio. Grape rootstocks were found to exclude chloride in saline soils and water.

Citrus fruits

Rangpur lime and Cleopatra mandarin rootstocks could withstand higher levels of sodium and chloride in irrigation water.

Leaf sampling technique was standardised for grapes, mango, banana, citrus, guava, pineapple, pomegranate, sapota, custard apple, papaya, passion fruit, jujube and phalsa for identifying the nutritional disorders.

A computer programme has been developed for fertiliser recommendations of different fruit crops. Best time of application and methods of placement of fertiliser have been worked out of sweet orange, Coorg mandarin, banana, Thompson Seedless grape and mango (Alphonso) based on root activity studies. Efficient citrus rootstocks of N,P and Zn absorption have been identified. Zn deficiency in sweet orange, complex bronzing in guava and B deficiency under high density planting in banana have been identified. Remedial measures have been recommended to rectify the disorders using micronutrients.

Vegetable Crops

Plant population density and optimum fertiliser requirements have been worked out in crops like tomato, egg plant, chilli, capsicum, cauliflower, cabbage, onion, okra, French bean and pumpkin.

Optimum agronomic practices like sequential-intercropping system, fertiliser and plant density levels for seed production, water requirement and optimum soil moisture regime for various crops have been worked out. Moisture stress at flowering was found critical for crop growth and yield in French bean (Contender). The middle node pods in okra are strong sinks for photoassimilates and could be used effectively for seed production. Spraying parachlorophenoxy acetic acid improved the fruit set (70%) and more size in Arka Saurabh, a heat sensitive tomato variety. Bioassay technique for herbicide residue detection has been standardised.

Salt tolerant lines in various vegetables, such as tomato, brinjal, okra, chillies and onion have been identified for use in marginal saline soils and water where electrical conductivity of the growing medium is between 2.5 and 5.0 dSm^{-1}.

Nutrient efficient (NPK) selections/ varieties in tomato, IIHR 550, IIHR 837, Arka Saurabh, Arka Vikas and Pusa Ruby can be directly utilised in the cropping programme as well as in marginal lands as they require less fertiliser and leave very little residue in the soil eliminating the hazards of salinity and pollution besides reducing the cost of fertiliser input. They can also be utilised in the breeding programme.

Mushrooms

A model mini bulk chamber was developed for composting and room pasteurisation methods were standardised. Shtick mushroom was successfully cultured for the first time in this part. Sporeless and low spore shredding strains of Pleurotus sp. have been developed. Studies on competitor molds associated with mushrooms and its control measures were worked out. The mushrooms laboratory is for the development of the industry in south India and assists by providing training, guidance, technological information and spawn.

Ornamental Crops

Pruning to 4-5 buds on the cane has yielded optimum flower production in rose. NPK 250:500:375 kg/ha was found to be optimum for flower production in rose planted closely. Planting gladiolus during the months of June and October-November yielded good quality spikes. Planting during the month of April-May was found to be the best in case of tuberose and planting of bulbs in flat bed was better than ridge planting. Protocorms of some orchids have been successfully encapsulated for storage and delivery and micro propagation techniques have been standardised. Micropropagation methods for anthuriums have been standardised for faster multiplication.

Medicinal and Aromatic Crops

Diosgenin assay procedure has been developed. Incubation techniques have resulted in obtaining higher diosgenin yield (16 to 58%) from Dioscorea floribunda tubers. Acid hydrolysis method of solasodine estimation has been developed. An increased yield of 8 to 12% essential oil from scented geranium was obtained by distilling the herbage after 17 hours but before 24 hours of harvest.

Crop Protection

Fruit crops

An Integrated Pest Management (IPM) strategy for control of fruit fly Bactrocera dorsalis on mango by a combination of cultural practices, spray at critical crop stages and post-harvest treatment of fruits in saline solution has been worked out. Biological control techniques were developed for the pink mealy bug, Maconellicoccus hirsutus on grapes with release of Australian lady bird beetle, Cryptolaemus montrouzieri and of citrus mealy bug (Planococcus citri) using the parasitoid, Leptomastix dactylopii. An integrated management of citrus nematode Tylenchulus semipenetrans in citrus has been developed. In grape, critical period of downy mildew disease has been identified to undertake effective and economical chemical control.

Biochemical factor for disease resistance has been worked out. In mango, sunshine hours (5 hours or less) play a major role in increasing powdery mildew disease. Several chemicals were found effective in checking the disease. For papaya and ber, powdery mildew, sunshine and relative humidity at 14.30 h had a profound impact on disease development. Extensive survey on guava wilt has been conducted in different parts of the country and five Fusarium spp., Botryciplodia sp. and Pestalotiopsis sp. have been reported. Post-harvest studies have established the occurrence of Pseudomonas sp., Enterobacter sp. and Penicillium sp. antagonistic to fungi causing fruit rot of guava. Treatments were also standardised to combat post-harvest and storage loss in banana and mango. The causal agent of Blastomania of citrus was identified as mycoplasma like organism (MLO). A bacilliform virus was discovered in pumello plants. Papaya ringspot virus was the most devastating on papaya and a wide variation was noticed among several serological groups in India. Copper resistance was noticed in field isolates of citrus canker pathogens.Cultural and chemical control were also standardised. Causal agent of pomegranate nodal blight has been confirmed as *X. campestris pv. punicae.*

Vegetables

Integrated pest management (IPM) strategies for major pest complex on cabbage and cauliflower with Indian mustard as a trap crop and African marigold as a trap crop for fruit borer (Helicoverpa armigera) on tomato have been developed. Optimum number of sprays for the management of beanfly and pod borers in leguminous crops as well as for fruit borer in okra and brinjal have been worked out. A sporeless mutant of Bacillus thuringiensis was isolated and utilised for the management of diamondback moth on crucifers. Nuclear polyhedrosis virus was isolated and used for the management of fruit borer on tomato. Root knot nematode Meloidogyne incognita in tomato and brinjal has been controlled by using oil cakes, endomycorrhizae and bio-agents like Trichoderma harzianum, Verticillium chlamydosporium and Paecilomyces lilacinus. Thai Sac brood disease of honey bees, which had brought down the number of bee colonies from 60 000 to 4000 was effectively controlled by inoculating an anti-viral agent.

A root-dip technique has been developed against Fusarium wilt of chilli and inoculation technique against Alternaria fruit rot of chilli and capsicum. An inoculation technique for screening against onion basal rot and root rot of pea have been developed. Sphaerotheca fulginer was found to be the most predominant among powdery mildew pathogens of cucurbits and control measures through chemical and cultural practices have been standardised for major diseases. Soil borne pathogens can be checked significantly by solarisation using polythene sheets to trap the heat change. Different MLO diseases of tomato and legumes have been identified. Sources of resistance to yellow vein mosaic of okra were identified. Molecular diagnostic technique for TLCV using DNA probe is developed for detecting TLCV in tomato and chilli. Fungicide combination of copper oxychloride and zinc based chemicals are effective for controlling the disease.

Memordicines from bitter gourd leaves showed antecedent activity against red pumpkin beetles and antifungal activity against Colletotrichum gloesporoides. The alkaloid serpentine present in roots of catharanthus showed in vitro nematicidal activity against root knot nematodes. Scented geranium oil and its constituent geraniol and citronella as well as methanol extract of onion seeds showed both antifungal activity against C. gloesporoides and nematicidal activity against root knot nematodes.

Ornamental Crops

Disease control methods were worked out for all important diseases of rose, gladiolus, chrysanthemum, jasmine, marigold, zinnia and crossandra. Isolates of Fusarium oxysporum f. sp. gladioli obtained from commercial farms of gladiolus were found to be Race-1 of F. oxysporum. Phetalaxyl was found to be phototoxic to crossandra among anti-oomycetes. Disease resistant sources were identified in rose and gladiolus. Effects of different domestic processes were studied to recommend the decontamination of residues from fruits and vegetables as worked out for use under domestic conditions. In vegetables, placement of fertiliser 5 cm below the seed/plant row showed maximum utilisation and as a result, 20% of saving in P input was achieved in tomato and onion and 40% in brinjal.

The persistence of various insecticides and fungicides on tropical fruits and vegetables was determined and safe waiting periods in terms of days required to be lapsed between the last application and harvest were determined, so these can be safely consumed without causing toxicity problems. The degradation of pesticide residues in soil following their build up during pest control operations was studied in different soils. Carrot and cauliflower responded to the application of B at the rate of 2 kg/ha as borax. Boron efficient lines in carrot and cauliflower have been identified.

Crop Utilisation

Handling and storage

Optimum maturity standards for mango, banana and tomato have been determined. Use of gibberellic acid and Vitamin K3, skin coating with Gal-prolong and Semperfresh enhanced the shelf life of mango by 2 to 4 days at 25^{0}C.

Individual shrink wrapping of capsicum, cucumber and pomegranate and storing at cool temperature (6^{0}C) showed a significant extension of shelf life upto 30 days. Use of ethylene absorbent in banana extended the shelf life both at 15^{0}C and 21^{0}C by 6 and 3 weeks respectively. Cultivars of grapes can be stored for 2-3 months at 1-3^{0}C using grape guard. Waxing of Coorg mandarins can extend shelf life when held at 25 and 10^{0}C. Pre-harvest nutritional status was observed to influence the post-harvest decay in onion.

Processing

Methods have been standardised for preparation of dehydrated products by freeze drying and osmotic dehydration from mango and banana. Technologies for raisin making, watermelon juice and banana beverage have been developed.

Microbiology

Methods have been standardised for making different types of wines. A method has been developed to make acceptable quality wine from high acid Bangalore Blue grapes. Vinegar with fruit flavour has been prepared from processing wastes. Methods for preservation of cut vegetables by lactic fermentation have been standardised.

ECONOMICS AND STATISTICS

The predominant role/expression of the female parent in crosses was established in 25 different crops. For the first time in the country, work on statistical ecology of insects was done and very useful information on the behaviour of the pest, its weakest point in life and sampling and forecasting methods were evolved. A novel method of finding out optimum plot size in plant protection trials was evolved for the first time. Generalised models for predicting fertiliser requirement of a crop based on available literature was built up in case of tomato. Using this fertiliser recommendations could be given village-wise. Expert system for guidance of cultivation of grapes and mushrooms was built up and supplied to the growers. Knowledge base of package of practices for 155 different horticultural crops was established for the four southern states. The cooperative marketing structure of horticultural crops was studied and many improvements for deriving better returns under efficient management were suggested. The working of cold storages were studied and economic inadequacies pointed out. The feasibility of cooperative small scale processing units in tomato and mango at village level was established.

Extension

During the last ten years, training programmes have been conducted at various levels to clientele such as state level officers of agricultural departments, university professionals and semi-government, voluntary organisations along with private enterpreneurs. In the last ten years, 2459 persons have been trained by this Institute on various aspects of horticultural production technologies through 128 training courses. Horticultural technologies released by the Institute were disseminated to officials of state departments, non-officials, farmers and entrepreneurs using variety of extension methods, which include in service training, organisation of field days, arranging exhibitions, advisory services to farmers, radio talks, TV talks, organisation of film and video shows, publication of extension bulletins and folders, field demonstration of new IIHR varieties.

Training of farmers

This is being done at the Krishi Vigyan Kendra at Gonoikoppal at Kodagu District. Around 8000 farmers, farm women and youths are trained in horticulture and related inputs every year by the KVK.

Advisory services

The soil and water samples received from farmers of Karnataka, Maharashtra and Tamil Nadu are analysed on payment basis and the nutrient status and fertiliser recommendations are sent to them for adoption.

Agricultural engineering

A continuous, motorised raw mango peeler of 200 kg h-1 capacity for pickle and chutney industries has been developed. By this, a saving of Rs.600/- can be effected when compared to manual peeling of one tonne. The time and manpower also is considerably saved. A fruit mechanised grader has also been developed for mango. Suitable packages for export of mangoes and pomegranate have also been developed.

Biotechnology

In a recalcitrant crop like mango, culture of nucellus and hybrid embryos from direct crosses yielded plantlets. Salt tolerant grape rootstock has been micro-propagated for its use on large scale. Hybrid grape variety Arka Neelmani has been multiplied in large scale for distribution to farmers. An efficient Sachet technique has been standardised for rapid acclimatisation of micropropagated grape.

Vesicular arbuscular mycorrhizal fungi and phosphate solubilising bacteria when inoculated to crops like papaya, banana and mango were found to substitute 50% urea and super phosphate in papaya cv. Coorg Honey Dew. 50% urea increased yield by 15-20%. In mango, cv. Arka Aruna, inoculation of VAM fungi improved plant growth. Nitrogen fixing bacteria colonising endorhizosphere of tomato, capsicum and brinjal have been isolated and their contribution to nitrogen economy has been found to be 20%. Bacteria isolated from tomato endorhizosphere has been identified to be a new species of Azospirillum. For the first time in India, a TLCV-in specific nucleic acid probe has been developed for gemini virus.

Expert Systems/Computer Aided Information

Considering the fact that computers are available in almost every district headquarters and also the fact that these programmes in Indian context can be used by extension personnel for solving the problems of cultivation, the Institute has initiated steps to use computers for transfer of technology in horticulture by developing expert system in various horticultural crops which can provide all the information about production technologies including propagation, cultivation, plant protection, package of practices and product utilisation. Work in this direction has successfully resulted in developing Grape Expert System, Mushroom Expert System, Cabbage Pest Expert System, Personal Computer based pest information on tomato and serpentine leaf miner information oriented computer aided system (SLM - INFO).

Apart from these, a new computer information system giving recommended guidelines on all aspects of cultivation of 153 horticultural crops (including plantation crops, tuber crops and spices) which gives recommended package of practices for each crop in Karnataka, Andhra Pradesh, Tamil Nadu and Kerala has been released. A data base programme giving guidance as to the fertiliser and spacing requirements of horticultural crops in India, statewise has also been developed to facilitate the extension worker, the planner or the cultivator to get within a few seconds the appropriate recommended doses of fertilisers, in whichever state he resides which will be released to general public for use shortly.

RESEARCH PREPAREDNESS FOR ACCELERATED GROWTH

The horticulture scenario of the country is rapidly changing. The production and productivity of horticultural crops have increased manifold. Production of fruits and vegetables has tripled in the last 50 years. The productivity has gone up by three times in banana and by 2.5 times in potato. Today horticultural crops cover about 25 per cent of total agricultural exports of the country. The corporate sector is also showing greater interest in horticulture. A major shift in consumption pattern of fresh and processed fruits and vegetables is expected in the coming century. There will be greater technology adoption both in traditional horticultural enterprise as well as in commercial horticulture sectors. Diversification and value addition will be the key words in the Indian horticulture in the 21st Century.

Horticulture research in India is about four decades old. Systematic research on fruit, vegetable and ornamental crops began in 1954 with the initiation of independent institutions and programmes. The research agenda is designed relevant to national plans and priorities for the horticulture development. Today, eight ICAR institutes with 27 regional stations, 1 project directorate, 10 national research centres, 16 all India coordinated research projects (AICRPS) with 223 research stations, 1 full-fledged university of horticulture, 25 state agricultural universities and 7 multi-disciplinary institutes of the ICAR are engaged in horticulture research. In addition, a few R&D establishments of crop/commodity boards and private sectors are providing research support to Indian horticulture. Research system in horticulture is now geared to provide necessary technological support to the expanding horticultural industry.

The research efforts in the past were mainly concentrated on crop improvement, propagation of seed/ planting material, agrotechniques, crop protection and post harvest management. Some of the improved technologies developed are enumerated below.

Technology Generation

Varietal development

Among the fruit crops, improved high yielding mango varieties, Mallika, Amrapali, Ratna, Sindhu, Arka Aruna, Arka Puneet, Dashehari-51 and hybrids namely CISH-M-1 and CISH-M-2 have been developed. Mallika is coming up in southern states like Karnataka and Amrapali is performing well in eastern India. Dashehari-51, a regular bearing cultivar with about 38 % higher productivity than the normal Dashehari, has been identified after 14 years of rigorous selection. In guava, three selections, namely Lalit, CISH-G-1 and CISH-G-2 have been developed for domestic and export markets. The fruits of Lalit are of medium size weighing about 150 g each and suitable for both table and processing purposes. In banana, high yielding hybrids like FHIA-01 and FHIA-03 are promising for replacing varieties, Panchananda and Bluggoe, respectively.

Cultivar, Saba is found promising under sodic soils. In addition, high yielding varieties like Col, Hl and H2 have been developed. In grape, superior and high yielding varieties have been developed *e.g.* Beauty Seedless and Pusa Seedless and early ripening variety, Perlette for cultivation under North Indian conditions and Anab-e-Shahi, Dilkhush, Thompson Seedless, Tas-A-Ganesh, Sonaka, Bangalore Blue and Pachadraksha for south Indian conditions. In citrus, high yielding and cluster bearing varieties of acid lime have been developed. *e.g.*, Rough Lemon, Rangpur Lime, Pramaini, Vikram PKM-1. Trifoliate oranges, namely Flying Dragon and Rich 16-6 are dwarfing types.

In papaya, high yielding superior varieties both for table purpose and papain production have been developed.

e.g., Co-1 to Co-7, Coorg Honey Dew, Pusa Delicious, Pusa Majesty, Pusa Giant and Pusa Nanha. In apple, superior hybrids have been developed. *e.g.*, Lal Ambari, Sunehari. Red Spur, Star Crimson, Golden Spur, Red Chief, Oregon Spur, Skyline Supreme and Vance Delicious have been identified. Tissue culture protocols for micro-propagation of two commercial varieties have been developed.

Among the vegetable crops, more than 130 open pollinated varieties, 36 hybrids, 3 synthetics and 29 resistant varieties of 20 vegetable crops have been developed and released for cultivation in different agro-climatic regions. These include 40 in tomato, 45 in brinjal, 13 in cauliflower, 12 in chillies, 20 in pea, 9 in musk melon, 16 in onion and 44 in other crops.

In potato, 33 high yielding varieties have been developed indigenously for large scale cultivation in different regions. Kufri Ashoka and Kufri Pukhraj mature in 75 days. Kufri Jawahar and Kufri Satluj are field resistant to late blight. Kufri Jawahar has most ideal plant type for inter-cropping. Kufri Chipsona-1 and Kufri Chipsona-2 have been developed with excellent processing attributes, comparable to exotic varieties. Kufri Swarna resistant to golden nematode is ideal for Nilgiri Hills.

In tuber crops, improved varieties of different tuber crops have been recommended/released for commercial cultivation. These includes 9 varieties of cassava, 15 varieties of sweet potato, 6 varieties of colocasia, 3 varieties each of greater yam and lesser yam, 1 each of *Amorphophallus*, taro and yambean. Cassava varieties, Sree Visakham and Sree Prakash are popular in Kerala. Triploid clone, Sree Harsha with high dry matter and starch content is suitable for industrial belt of Tamil Nadu. Two early maturing varieties Sree Jaya and Sree Vijya have been released for culinary purposes. Elephant foot yam variety, Am-15 has been released with high yield potential of 41 t/ha.

Among the plantation and spice crops, India is the first country to exploit hybrid vigour in coconut. Twelve hybrids involving tall and dwarf parents and 4 varieties have been released for commercial cultivation. These varieties yield 21 to 89 % more than the local cultivars. Some of the released varieties like Chandra Kalpa and Pratap (Banawali Green Round) are receiving wide acceptance of farmers. Chowghat Green Dwarf variety is good for tender coconut purpose. In arecanut, 4 high yielding varieties, namely, Mangala, Sumangala, Sreemangala and Mohitnagar have been developed, giving about 30 % higher yield than the local cultivars.

In oil palm, first efforts for improvement were made by producing Tenera hybrids using Pisifera pollen imported from Nigeria. Dura x Pisifera hybrids are field tested in East and West Godavari districts, Khammam and Krishna districts of Andhra Pradesh, with yield potential of 20-25 tonnes/ha of FFB from the fifth year. In cashew, 22 region specific selections and 12 hybrids with yield potential of 1.5-2 tonnes of raw nuts/ha have been produced and released for commercial cultivation. The present standards fixed for cashew varieties include export grade kernels of W-210 to W-240 and at least one tonne per ha yield with 30 per cent shelling. In black pepper, 6 varieties, namely, Sreekara, Subhakara, Palode-2, Panniyur-2, Panniyur-4, Panchami and Pournami, and 2 hybrids *viz.* Panniyur-1 and Panniyur-4 have been developed.

In cardamom, a number of improved varieties have been developed and released for commercial cultivation *e.g.*, CCS-1, Mudigere-1, PV-1, ICRI-1 and ICRI-2. In ginger, varieties like Suprabha, Suruchi, Suravi and Varada have been developed. In turmeric, several varieties *viz.*, Co-1, Krishna, Sugandham, BSR-1, Suvarna, Roma, Suroma, Rajendra Sonia, Sugana, Sudarshana, Ranga, Rasmi, Prabha, Prathiba, Mega Turmeric and RCT-1 have been developed with yield potential of up to 44 tonnes of fresh rhizomes per ha. Three high yielding cinnamon lines, namely, Navashree, Nithyashree and Konkan Tej have been released for cultivation.

Seed/planting Material Propagation

In most of the fruit crops, vegetative propagation techniques have been standardized. Soft wood grafting has been standardized for mango, sapota, custard-apple and jackfruit. Other vegetative propagation techniques have been developed for *ber, aonla,* jackfruit, custard-apple and *bael*. In mango, veneer grafting and stone-grafting is practised commercially. Mango variety, Vellaikolumban is suitable semi-dwarfing rootstock for Alphonso. Old unproductive mango trees can be rejuvenated successfully by pruning the 4th order branches during December-January. Flowering and fruiting are regular in pruned trees. For mandarin orange, Rangpur lime is a drought hardy rootstock. In grapes, Dogridge and Salt Creek (Ramsey) are suitable for minimizing adverse effects of soil salinity on Thompson Seedless.

A tissue culture technique for mass multiplication of Dogridge has been standardised. In banana, sword suckers of 700-1000 g are optimum. Rhizomes with active lateral buds and dead central buds are preferred for distant transportation in western India. Double paring and shade drying followed by dipping in Monocrotophos (0.5 %) and Bavistin (0.2 %) is recommended to disinfect nematodes and soil borne fungi. Tissue cultured banana plants are now commercially adopted for their uniformity in flowering and produce. Shoot-tip grafting technique in citrus has been considerably advanced.

In vegetable crops, seed production of over 120 open pollinated high yielding varieties of different vegetables has been well established in the country. Hybrid seed production has become easier with the development of male sterile lines in tomato, self incompatible lines in cauliflower and gynoecious lines in cucumber and muskmelon. In brinjal, functional male sterility controlled by a single recessive gene has been identified. Temperature barrier in cole crops (cabbage and cauliflower) has been overcome by developing heat tolerant hybrids. It is now possible to cultivate cabbage and cauliflower in southern India. Development of tomato varieties resistant to bacterial wilt has made their cultivation successful in non-traditional areas. Onion seed production technology for cultivation in kharif season has been developed for north Indian states especially, Haryana, Punjab and western Uttar Pradesh.

Seed Plot Technique has been developed for production of disease-free potato seed in plains. It is widely adopted by farmers. A new technology for raising commercial crop of potato using 'True Potato Seed' (TPS) has been developed and standardized as supplementary technology to the traditional tuber grown crop. Two TPS populations, TPS-C-3 and HPS-113 are recommended for commercial production in Bihar, Gujarat, Tripura and West Bengal. Micropropagation protocols have been developed in banana, oil palm, cashew, black pepper, ginger, *etc.* Seed gardens of Tall (T) x Dwarf (D) and D x T hybrids have been established for production of coconut hybrids.

Agrotechniques

In fruit crops, improved agrotechniques developed have helped the farmers in improving the productivity and quality of produce. Soil application of paclobutrazol (4 g/ tree) increase flowering and fruiting in mango on commercial scale in coastal Maharashtra. It also controls irregular bearing in cultivar Dashehari. Spray of NAA @ 200 ppm in October is recommended for control of malformation. Heavy fruit drop at maturity in cultivar Langra can be controlled by spraying NAA (20 ppm). In guava, double spray of 10 and 20 % urea on cultivars, Allahabad Safeda and Sardar twice at bloom eliminate poor quality rainy season crop and increases winter season yield by 3 and 4 times, respectively.

Application of neem coated urea (800 g/plants) yields 98 kg fruits/plant in variety, Sardar compared to 37 kg from untreated plants. In banana, high density planting (4550 plants/ha) yield up to 174 t/ha. Adoption of improved technology in Maharashtra has resulted in fruit yield increase up to 52 t/ha. In citrus, two grafting methods using inverted 'T' cut and apical triangle cut have been developed with overall success of around 36% using either method.

For accelerating the survival of growth, shoot tip grafts, the successful grafts are double grafted (side grafted) on vigorous green-house grown Rough Lemon and Rangpur Lime seedlings. Rangpur Lime rootstock is found superior for sweet oranges and mandarins. In papaya, closer spacing of 1.4 x 1.4 m is recommended for high yield. Drip irrigation techniques have been standardized. In banana, it has resulted in production gain (60-70%) and early harvesting (40-50 days), besides improved water efficiency. Likewise, in grapes higher yields have been obtained with better water use efficiency (11 %).

In vegetable crops, improved production technology has been developed for major crops. Drip irrigation is economical in tomato and brinjal. In cucumber, replenishment of evaporation loss through irrigation resulted in maximization of yield of quality fruits. Drip irrigation in watermelon provided 33% higher yield with a water saving of 40%. Nutrient requirements and fertiliser schedules have been worked out crop-wise and recommended for different agro-climatic regions. In leguminous vegetables, high N depresses nodulation. The VAM fungi increases P availability to plants.

In all leguminous vegetables, inoculation of the VAM fungi along with *Rhizobium* culture is beneficial. Production technologies for *kharif* onion in northern India and long day type onions for high altitudes have been standardized.

Pendimethalin (Stomp) has been found effective in controlling weeds in tomato, brinjal, chilli, bell pepper and okra. In potato, a number of potato-based multiple and inter cropping systems have been developed for different potato growing regions. Intercrop combinations with sugarcane in Maharashtra, wheat in Chhota Nagpur area and linseed in central Uttar Pradesh are found remunerative. A suitable method of urea application has also been worked out.

In tuber crops, short duration legumes *viz.*, groundnut and French bean and cowpea can be successfully inter-cropped with cassava. Short-duration cassava, Sree Prakash is ideal in double cropped rice fields. Studies on cassava-based multiple cropping systems involving banana, coconut, *Leucaena* and Eucalyptus, have shown banana-cassava combination to give maximum root yield. Banana and coconut combination reduces soil loss and surface run-off considerably. *Dioscorea* and elephant foot yam with banana, Nendran can generate an additional income of Rs 20,000/- over the sole crop of banana. For Inter-cropping *Dioscorea* in coconut garden, the ideal planting density is 9000 plants/ha. When *Amorphophallus* is raised as an inter-crop in coconut garden, one third dose of recommended fertilizers is sufficient. Inoculation with VAM fungi in cassava give about 15-20% increase in yield.

Among the plantation and spice crops, density of 175 coconut palms/ha (7.5 m x 7.5 m spacing) is found ideal. In general, NPK application of 500:320:1200 g/palm/year is found optimum. A multi-storied cropping system involving black pepper trained on coconut trees, and cocoa in between the rows of coconut and pineapple in the ground floor has been found ideal for exploiting light, soil and air spaces. In arecanut, application of NPK (100:40:140 g) and green leaf (14 kg) per palm per year is recommended for coastal regions of Kerala and Karnataka and for plains of West Bengal, Karnataka and Assam.

In oil palm, application of NPK (1200:600:1200 g/palm) is found to give 17.1 tonne of FFB per ha. In black pepper, rapid methods for production of rooted cuttings of pepper have been developed and a commercial protocol has been standardised for micropropagation of black pepper. Ginger yield could be increased up to 33 percent by application of neem cake at the rate of two tonnes per ha and the fertiliser schedule of 75 kg each of N, P205 and K20. Technology for storage of ginger seed rhizome is standardised and recommended.

Protection Technologies

For major fruit crops, plant protection schedules have been developed for the control of significant insect-pests for wider adoption. In the recent years, research efforts are directed to devise eco-friendly, economical and long lasting control measures. Success has been achieved in biological control of mealy bugs in mango and guava. The *Beauveria bassiana* has been found killing mango mealy bug and hopper. In grapes, integrated management of *Spodoptera* caterpillar involving light and pheromone traps, NPV and neem based insecticides and biological control of mealy bug by the beetle *Cryptolaemus montrouzieri* have been standardized. Studies on pesticide residues have resulted in working out of safe-waiting periods for harvesting and consumption of fruits.

In vegetable crops, about 50 improved measures for efficient management of diseases and 23 for insect-pests have been worked out and popularized in different agro-climatic regions in the country. Integrated pest management (IPM) for controlling diamond back moth on cabbage through a trap crop like mustard has been demonstrated. Fruit borer *(H. armigera)* on tomato can be controlled by the release of *Trichogramma pretiosum* alone and in combination with HaNPV.

In potato, integrated management schedules for control of bacterial wilt and tuber moth have been developed. A late blight forecasting system has been developed for the hills.

Among the tuber crops, the major diseases affecting tuber crops are cassava mosaic and brown leaf spot in cassava, *Phytophthora* leaf blight in colocasia, *Fusarium* wilt in elephant yam and virus diseases of sweet potato. Foliar sprays of Bavistin (0.1%) combined with disodium and dipotassium phosphates (100 ppm) and calcium sulphate at 15-day interval was found to check foliar diseases in sweet potato. The major pests include spider mites, scale insects and white fly on cassava, weevil on sweet potato, defoliators, aphids and mites on colocasia, and scales and mealy bugs on yams and elephant yam. Cultural methods for weevil control include clean cultivation, destruction of alternate hosts and timely harvest. An effective IPM package using synthetic sex pheromone has been developed. Control measures involving insecticides have been evolved for the control of pests of other tuber crops.

Among the plantation and spice crops, bud rot of coconut caused by *Phytophthora palmivora* can be effectively controlled by spraying Bordeaux mixture. Calyxin root feeding and drenching of soil with 1% Bordeaux mixture along with neem cake application @ 5 kg per palm per year is recommended for controlling Thanjavur wilt disease reported in Tamil Nadu, Andhra Pradesh and Karnataka. A package of practices has been developed for managing mycoplasma like organisms (MLOs) in root wilt affected coconut palms in Kerala and Thatipaka disease affected palms in Andhra Pradesh.

Eradication of all root wilt affected palms is recommended. In cashew, tea mosquito bug (TMB) can be effectively controlled through a schedule of spray coinciding with flushing, flowering and fruiting. For effective control of stem and root borer infestation, constant monitoring and adoption of strict sanitation in the plantations coupled with prophylactic application of coal tar and kerosene in the ratio of 1:2 on trunks are recommended. In black pepper, spraying Bordeaux mixture (1 %) and drenching the soil with copper oxychloride (0.2 %) is found effective in managing *Phytophthora* foot rot.

Post Harvest Management Technologies

The post harvest handling of fruits and vegetables accounts for 20-30% of losses at different stages of storage, grading, packing, transport and finally at marketing as a fresh produce or in processed form. A number of improved technologies have been developed for commercial exploitation.

An on-farm, low cost, environment friendly cool chamber, Zero Energy Cool Chamber has been developed using locally available material. The principle of evaporative cooling reduces the inside temperature by as much as 17-18 oC and keeps the relative humidity above 90% during peak summer. The chamber increases the shelf life and reduces PLW of banana, mango, orange lime, grape fruit, tomato and potato in different situations in India.

Maturity standards for mango, guava, grape, litchi and *ber* and chemical treatments for regulation of ripening in mango, sapota and banana have been standardised. Optimum storage temperatures worked out for several fruits, vegetables and tuber crops.

A mango harvester, fruit peeler, hand and pedal operator cassava chipping machines, harvesting tools (5-14 times efficient), coconut dehusking machine, implements for mechanization of potato cultivation and other crops have been developed.

A number of improved technologies have been developed for commercial exploitation *viz.*, tent type foldable solar dryer, packaging boxes for distant transportation of apple, mango, citrus and plum, production of value added products-pectin from peel and flour from mango fruit kernel, production of fruit post carbonated beverages *etc.*

Production Constraints

In spite of great strides made, the productivity of horticultural crops, in general, is still quite low and the post harvest losses particularly of perishable commodities, are considerable. Improvement in quality standards of the produce and their marketing are essential to increase our share in the global market.

The research agenda in horticulture is by design relevant to national plans and priorities and research programmes are normally formulated keeping in view the thrust areas in development. The major technology related constraints contributing to low productivity of horticultural crops and inferior quality of produce are:

Vast majority of holdings are small and un-irrigated. Large tracts of low and unproductive plantations needing replacement/rejuvenation. Low productivity of crops due to inferior genetic stocks and poor management. Inadequate supply of quality planting materials of improved varieties. High incidence of pests and diseases. Heavy post harvest losses and low utilization in processing sector.

For addressing the above constraints, research institutions are engaged in both basic and applied research. While formulating research strategies some of the inherent weaknesses associated with perennial tree crops and certain perpetual problems in Indian horticulture must be kept in mind. They are:

Long period required for development of improved genotypes. Application of biotechnological tools/methods in horticultural crops is still in its early stage of development in the country. Chronic production problems due to major disorders like alternate bearing, malformation and spongy tissue in mango, guava wilt, citrus decline, root wilt in coconut, viral disease in vegetables, *Phytophthora* diseases in large number of crops *etc.* remained largely unresolved. Lack of advanced technologies for post harvest handling, processing and marketing of produce.

Losses caused by biotic stresses are very high and due to pesticide residue problems development of eco-friendly IPM strategy is more relevant in horticulture. There is a threat for loss of valuable genetic resources, if measures are not taken for their conservation. Wastelands and hilly terrains being the potential future expansion areas, matching technologies for dry land and hill horticulture need to be developed. Counter seasonal advantages from diverse agro-climatic situations provide strength for extended availability of horticultural crops round the year and such potentials can be harnessed only with relevant research support.

Research Strategies and Programmes

Keeping in view of the strength and weaknesses of the research system and priority areas of horticulture development the following strategies and programmes are suggested for horticultural research:

Rationalization of Research

There is a need to shift from commodity/discipline oriented research to system based research and to establish stronger Inter-institutional linkages with SAUs, CSIR, DBT and BARC. Greater private sector partnership for diversification, value addition and export promotional research and seed production programmes will be required to modernize horticulture industry. Introduction of project based budgeting in the Institutes/NRCs will bring better accountability and one time catch up grant to modernize old Institutes will be essential. Development of database on technologies evolved, market intelligence, export projections/removal of quantitative restrictions, R&D scenario in horticulture should receive priority attention.

Safeguard for Intellectual Property Rights (IPR)

A fool proof description of varieties and their registration and finalisation of material transfer agreement and channelization of germplasm exchange need to be institutionalized. Similarly, specification of quality/codex standard for export of indigenous fruits and vegetables need to be developed. Phytosanitary regulations for importing vegetatively propagated materials need a relook/revision and rigorous enforcement.

Genetic Resource Management

Greater emphasis need to be given on (a) *in situ* conservation of endangered species, *ex situ* conservation of base collections and *in vitro* storage and cryopreservation of important germplasm. Clonal repository of vegetatively propagated crops and germplasm screening for processing and diversified use will be certain other areas of priority research.

Horticultural Biotechnology

Biotechnology as a tool for rapid multiplication of quality planting material, virus cleaning, genetic transformation *etc.*, will be of very great importance. The priority research programmes are:

- — Development of micropropagation protocols in selected crops
- — Genetic engineering for integration of desirable traits
- — Molecular characterization of germplasm and development of molecular linkage maps

— Value addition to products
— Preservation of post harvest losses through control of metabolic process

Integrated Production System

Production related technologies can bring quick improvement in production and productivity on different regions. Shor or medium range programme on horticulture based cropping systems; water management including microirrigation and fertigation, greenhouse cultivation of vegetables and flowers, integrated nutrient and pest managements, environment pollution and pesticide residue problems have already received research attention. Further refinement of the technologies and their transfer will bring perceptible change/improvement in production of different crop commodities.

Export Promotional Research

There is a need for development of bulk handling system of tropical fruits, including pre-cooling and CA/MA storage and post harvest protocols for sea transport of major fruits like banana, mango, litchi, sapota, Kinnow and pomegranate. Disinfestation technology including vapour heat treatment (VHT) for export of fresh fruits and extension of shelf life by preventing desiccation of vegetables should help in further export promotion. Organic farming for vegetable and spice crops and residue free IPM technology are other important areas of research.

Quality Planting Material Production

For resolving the long standing problems of supply of good quality planting materials of different horticultural crops, following research areas are flagged:

System of certification and standardisation of planting materials; Micropropagation protocols for mango, litchi, coconut, walnut, date palm, oil palm and apple rootstocks; Refinement of TPS and micro-tuber production technology in potato; Improvement in STG techniques and cross-protection in citrus for virus elimination and control; and Seed production system for hybrid vegetables and commercialization of micropropagation in floriculture.

Post Harvest Management

In order to reduce post harvest losses at production centres low cost ecofriendly on-farm storage structures can play a crucial role. Significant advancement has been made in that direction and some small and medium sized cool chambers on the principles of evaporative cooling have been devised. Further refinement of the technology will go on a long way. Also, standardisation of packing line operations and proper packaging of different commodities are of urgent need. Pesticide residue management and newer product developments will add values to the produce.

Human Resource Development

Advanced training in research methodologies and instrumentation, biotechnology, micro-irrigation, fertigation, IPM, INM, biofertilizer, biopesticide, pesticide residue, PHT and product development need priority attention for increasing research capabilities of the scientists. Skill development for state level development functionaries through in-service training at different R&D institutions will enhance capabilities of extension staff. Post Graduate programmes in fruit, vegetable, floriculture, plantation crops and post harvest management of horticultural crops will help in providing trained manpower in specialized areas.

Technology adoption pattern can not be uniform throughout the country and will vary from crop to crop and even from region to region. Certain degree of flexibility in research planning and research strategy is therefore obvious. Also, with the opening of global markets and removal of quantitative restrictions under the WTO export-import scenario is likely to change at much faster pace. Market forces will play a more dominant role and demands for modern technologies will increase. Research system in horticulture will have to be very alert and should be able to adjust with the changes. Development of both short term and long term strategies for modernising Indian horticulture will depend largely on the research support and strength of research system.

REFERENCES

Bantilan, M.C.S. and P.K. Joshi, "Returns to research and diffusion investments on wilt resistance in pigeonpea", *Impact Series 1*, Patancheru, India: ICRISAT, 1996.

Singh, R.P. and M. L. Morris, "Adoption, management and impact of hybrid maize seed in India", *CIMMYT Economics Program Working Paper*, Mexico, D.F.: CIMMYT, 1997.

Kloppenburg, J., "Social Theory and the De/Reconstruction of Agricultural Science: Local Knowledge for an Alternative Agriculture", *Rural Sociology,* 1991.

Krell, Rainer, FAO (ed.), "Biological Farming Research in Europe", *REU Technical Series No. 54,* FAO Regional Office for Europe, Rome, 1997.

Kuhn, T .S., *The structure of scientific revolutions,* 3. Ed.Chicago: The University of Chicago Press, 1996.

Kumar, P., C.C. Maji and R.K. Patel, "Returns on Investment in Research and Extension: A study on Indo-Swiss Cattle Improvement Project, Kerala", Indian Journal of Agricultural Economics, 1977.

Maredia, M.K. and D. Byerlee, "Efficiency of research investments in the presence of international spillovers: Wheat research in developing countries", *Agricultural Economics,* 2000.

Niggli, U. and Lockeretz, W., "Development of Research in Organic Agriculture", In: Oestergaard, (Ed.): *Fundamentals of Organic Agriculture,* IFOAM, Tholey-Theley, 1996.

Oehmke, J. and E. Crawford, "The impact of agricultural technology in subsaharan Africa,""*Journal of African Economies,* 1996.

Omiti, J., M. Waithaka, and W. Mwangi, "Situation of impact assessment of agricultural research in the ASARECA region", *Paper prepared for the workshop on impact assessment of agricultural research in the East and Central Africa,* Entebbe, Uganda, November 16-19,1999.

Ozgediz, S., "Evaluating research institutions: lessons from the CGIAR", In *Knowledge, Technology and Policy.* Vol. 11,1999.

Paarlberg, R.L., "External impact assessment of IFPRI's 2020 Vision for food, agriculture, and the environment initiative", *IFPRI Impact Assessment Discussion Paper 10,* Washington, D.C.: IFPRI, 1999.

Poesch, G. H., and Laurie, Alex, "The use of artificial light and reduction of the daylight period for flowering plants in the greenhouse", *Ohio Agr. Exp. Sta. Bull.,*1935.

Rosenau, P.M., *Post-Modernism and the Social Sciences: Insights, Inroads and Intrusions,* Princeton, N.J: Princeton University Press, 1992.

Singh, R.P. and M. L. Morris, "Adoption, management and impact of hybrid maize seed in India", *CIMMYT Economics Program Working Paper,* Mexico, D.F.: CIMMYT, 1997.

Index